MW01309094

INNOVATIVE SITE REMEDIATION TECHNOLOGY: DESIGN AND APPLICATION

CHEMICAL TREATMENT

One of a Seven-Volume Series

Prepared by WASTECH®, a multiorganization cooperative project managed by the American Academy of Environmental Engineers® with grant assistance from the U.S. Environmental Protection Agency, the U.S. Department of Defense, and the U.S. Department of Energy.

The following organizations participated in the preparation and review of this volume:

Air & Waste Management Association
P.O. Box 2861
Pittsburgh, PA 15230

American Society of Civil Engineers
345 East 47th Street
New York, NY 10017

American Academy of Environmental Engineers®
130 Holiday Court, Suite 100
Annapolis, MD 21401

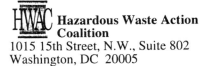

Hazardous Waste Action Coalition
1015 15th Street, N.W., Suite 802
Washington, DC 20005

American Institute of Chemical Engineers
345 East 47th Street
New York, NY 10017

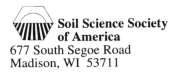

Soil Science Society of America
677 South Segoe Road
Madison, WI 53711

Water Environment Federation
601 Wythe Street
Alexandria, VA 22314

Monograph Principal Authors:
Leo Weitzman, Ph.D., *Chair*
Irvin A. Jefcoat, Ph.D. **Byung R. Kim, Ph.D.**

Series Editor
William C. Anderson, P.E., DEE

Library of Congress Cataloging in Publication Data

Innovative site remediation technology: design and application.
p. cm.
"Principle authors: Leo Weitzman, Irvin A. Jefcoat, Byung R. Kim"--V.2, p. iii.
"Prepared by WASTECH."
Includes bibliographic references.
Contents: --[2] Chemical treatment
1. Soil remediation--Technological innovations. 2. Hazardous waste site remediation--Technological innovations. I. Weitzman, Leo. II. Jefcoat, Irvin A. (Irvin Atly) III. Kim, B.R. IV. WASTECH (Project)
TD878.I55 1997
628.5'5--dc21 97-14812
CIP

ISBN 1-883767-17-2 (v. 1)
ISBN 1-883767-18-0 (v. 2)
ISBN 1-883767-19-9 (v. 3)
ISBN 1-883767-20-2 (v. 4)
ISBN 1-883767-21-0 (v. 5)
ISBN 1-883767-22-9 (v. 6)
ISBN 1-883767-23-7 (v. 7)

Copyright 1997 by American Academy of Environmental Engineers. All Rights Reserved. Printed in the United States of America. Except as permitted under the United States Copyright Act of 1976, no part of this publication may be reproduced or distributed in any form or means, or stored in a database or retrieval system, without the prior written permission of the American Academy of Environmental Engineers.

The material presented in this publication has been prepared in accordance with generally recognized engineering principles and practices and is for general information only. This information should not be used without first securing competent advice with respect to its suitability for any general or specific application.

The contents of this publication are not intended to be and should not be construed as a standard of the American Academy of Environmental Engineers or of any of the associated organizations mentioned in this publication and are not intended for use as a reference in purchase specifications, contracts, regulations, statutes, or any other legal document.

No reference made in this publication to any specific method, product, process, or service constitutes or implies an endorsement, recommendation, or warranty thereof by the American Academy of Environmental Engineers or any such associated organization.

Neither the American Academy of Environmental Engineers nor any of such associated organizations or authors makes any representation or warranty of any kind, whether express or implied, concerning the accuracy, suitability, or utility of any information published herein and neither the American Academy of Environmental Engineers nor any such associated organization or author shall be responsible for any errors, omissions, or damages arising out of use of this information.

Printed in the United States of America.

WASTECH and the American Academy of Environmental Engineers are trademarks of the American Academy of Environmental Engineers registered with the U.S. Patent and Trademark Office.

CONTRIBUTORS

PRINCIPAL AUTHORS

Leo Weitzman, Ph.D., *Task Group Chair*
President
LVW Associates, Inc.

Irvin A. Jefcoat, Ph.D.
University of Alabama
Department of Civil Engineering

Byung R. Kim, Ph.D.
Ford Motor Research Laboratory
Energy Systems Division

REVIEWERS

The panel that reviewed the monograph under the auspices of the Project Steering Committee was composed of:

Richard A. Conway, P.E., DEE, *Chair*
Union Carbide Corporation

Herbert E. Allen, Ph.D.
University of Delaware

William J. Cooper, Ph.D.
High Voltage Environmental Applications, Inc.

Ernest Gloyna, Ph.D., P.E., DEE
University of Texas

Eric Lindgren, Ph.D.
Sandia National Laboratories

Steven McCutcheon, Ph.D., P.E.
EPA — Athens, Georgia

Michael G. Nickelsen, Ph.D.
High Voltage Environmental Applications, Inc.

Steven Shoemaker, P.E.
DuPont Environmental

Michael H. Spritzer
General Atomics

Walter J. Weber, Jr., Ph.D., P.E., DEE
University of Michigan

STEERING COMMITTEE

This monograph was prepared under the supervision of the WASTECH® Steering Committee. The manuscript for the monograph was written by a task group of experts in chemical treatment and was, in turn, subjected to two peer reviews. One review was conducted under the auspices of the Steering Committee and the second by professional and technical organizations having substantial interest in the subject.

Frederick G. Pohland, Ph.D., P.E., DEE *Chair*
Weidlein Professor of Environmental Engineering
University of Pittsburgh

Richard A. Conway, P.E., DEE, *Vice Chair*
Senior Corporate Fellow
Union Carbide Corporation

William C. Anderson, P.E., DEE
Project Manager
Executive Director
American Academy of Environmental Engineers

Colonel Frederick Boecher
U.S. Army Environmental Center
Representing American Society of Civil Engineers

Clyde J. Dial, P.E., DEE
Manager, Cincinnati Office
SAIC
Representing American Academy of Environmental Engineers

Timothy B. Holbrook, P.E.
Engineering Manager
Camp Dresser & McKee, Incorporated
Representing Air & Waste Management Association

Joseph F. Lagnese, Jr., P.E., DEE
Private Consultant
Representing Water Environment Federation

Peter B. Lederman, Ph.D., P.E., DEE, P.P.
Center for Env. Engineering & Science
New Jersey Institute of Technology
Representing American Institute of Chemical Engineers

George O'Connor, Ph.D.
University of Florida
Representing Soil Science Society of America

George Pierce, Ph.D.
Manager, Bioremediation Technology Dev.
American Cyanamid Company
Representing the Society of Industrial Microbiology

Peter W. Tunnicliffe, P.E., DEE
Senior Vice President
Camp Dresser & McKee, Incorporated
Representing Hazardous Waste Action Coalition

Charles O. Velzy, P.E., DEE
Private Consultant
Representing, American Society of Mechanical Engineers

Calvin H. Ward, Ph.D.
Foyt Family Chair of Engineering
Rice University
At-large representative

Walter J. Weber, Jr., Ph.D., P.E., DEE
Gordon Fair and Earnest Boyce Distinguished Professor
University of Michigan
Representing Hazardous Waste Research Centers

FEDERAL REPRESENTATION

Walter W. Kovalick, Jr., Ph.D.
Director, Technology Innovation Office
U.S. Environmental Protection Agency

George Kamp
Cape Martin Energy Systems
U.S. Department of Energy

Jeffrey Marqusee
Office of the Under Secretary of Defense
U.S. Department of Defense

Timothy Oppelt
Director, Risk Reduction Engineering Laboratory
U.S. Environmental Protection Agency

REVIEWING ORGANIZATIONS

The following organizations contributed to the monograph's review and acceptance by the professional community. The review process employed by each organization is described in its acceptance statement. Individual reviewers are, or are not, listed according to the instructions of each organization.

Air & Waste Management Association

The Air & Waste Management Association is a nonprofit technical and educational organization with more than 14,000 members in more than fifty countries. Founded in 1907, the Association provides a neutral forum where all viewpoints of an environmental management issue (technical, scientific, economic, social, political, and public health) receive equal consideration.

Qualified reviewers were recruited from the Waste Group of the Technical Council. It was determined that the monograph is technically sound and publication is endorsed.

The reviewers were:

Lee Dodge
RUST Remediation Services
Pleasanton, CA

James Strunk
Union Carbide Corporation
Bound Brook, NJ

American Institute of Chemical Engineers

The Environmental Division of the American Institute of Chemical Engineers has enlisted its members to review the monograph. Based on that review the Environmental Division endorses the publication of the monograph.

American Society of Civil Engineers

The American Society of Civil Engineers, established in 1852, is the premier civil engineering association in the world with over 124,000 members. Qualified reviewers were recruited from its Environmental Engineering division. These individuals reviewed the monograph and have determined that it is acceptable for publication.

The reviewers were:

G. Fred Lee, Ph.D., P.E., DEE
G. Fred Lee & Associates
El Macero, CA

Richard Reis, P.E.
EMCON
Bothell, WA

Hazardous Waste Action Coalition

The Hazardous Waste Action Coalition (HWAC) is the premier business trade group serving and representing the leading engineering and science firms in the environmental management and remediation industry. HWAC's mission is to serve and promote the interests of engineering and science firms practicing in multi-media environment management and remediation. Qualified reviewers were recruited from HWAC's Technical Practices Committee. HWAC is pleased to endorse the monograph as technically sound.

The lead reviewer was:

James D. Knauss, Ph.D.

President, Shield Environmental

Lexington, KY

Soil Science Society of America

The Soil Science Society of America, headquartered in Madison, Wisconsin, is home to more than 5,300 professionals dedicated to the advancement of soil science. Established in 1936, SSSA has members in more than 100 countries. The Society is composed of eleven divisions, covering subjects from the basic sciences of physics and chemistry through soils in relation to crop production, environmental quality, ecosystem sustainability, waste management and recycling, bioremediation, and wise land use.

Members of SSSA have reviewed the monograph and have determined that it is acceptable for publication.

The lead reviewer was:

John Zachara, Ph.D.

Pacific Northwest Laboratory

Richland, WA

Water Environment Federation

The Water Environment Federation is a nonprofit, educational organization composed of member and affiliated associations throughout the world. Since 1928, the Federation has represented water quality specialists including engineers, scientists, government officials, industrial and municipal treatment plant operators, chemists, students, academic and equipment manufacturers, and distributors.

Qualified reviewers were recruited from the Federation's Hazardous Wastes Committee and from the general membership. It has been determined that the document is technically sound and publication is endorsed.

The reviewers were:

James A. Kent, Ph.D.

Morgantown, WV

William Butler

DuPont Environmental Remediation Services

Wilmington, DE

ACKNOWLEDGMENTS

The WASTECH® project was conducted under a cooperative agreement between the American Academy of Environmental Engineers® and the Office of Solid Waste and Emergency Response, U.S. Environmental Protection Agency. The substantial assistance of the staff of the Technology Innovation Office was invaluable.

Financial support was provided by the U.S. Environmental Protection Agency, Department of Defense, Department of Energy, and the American Academy of Environmental Engineers®.

This multiorganization effort involving a large number of diverse professionals and substantial effort in coordinating meetings, facilitating communications, and editing and preparing multiple drafts was made possible by a dedicated staff provided by the American Academy of Environmental Engineers® consisting of:

William C. Anderson, P.E., DEE
Project Manager & Editor

John M. Buterbaugh
Assistant Project Manager & Managing Editor

Robert Ryan
Editor

Catherine L. Schultz
Yolanda Y. Moulden
Project Staff Production

J. Sammi Olmo
I. Patricia Violette
Project Staff Assistants

TABLE OF CONTENTS

LIST OF TABLES

LIST OF FIGURES

1 INTRODUCTION

This monograph, covering the design and applications of Chemical Treatment, is one of a series of seven on innovative site and waste remediation technologies. This series was preceded by eight volumes published in 1994 and 1995 covering the description, evaluation, and limitations of the processes. The entire project is the result of a multiorganization effort involving more than 100 experts. It provides the experienced, practicing professional guidance on the innovative processes considered ready for full-scale application. Other monographs in this design and application series and the companion series address bioremediation, liquid extraction: soil washing, soil flushing, and solvent/chemical, stabilization/solidification, thermal desorption, thermal destruction, and vapor extraction and air sparging.

1.1 Background

An earlier book on chemical treatment (Weitzman et al. 1994) categorizes the technology into three processes:

> *Substitution Processes* that substitute a different functional group for one or more functional groups on a target molecule. For example, a mixture of potassium hydroxide and polyethylene glycol (PEG) is used to replace one or more chlorine atoms on a polychlorinated biphenyl (PCB) molecule with a PEG moiety. The resulting molecule is not legally a PCB nor is it regulated by the Toxic Substances Control Act. The hazard of the PEG moiety is unknown.
>
> *Oxidation Processes* that use an oxidizing agent, such as air, oxygen, ozone, or hydrogen peroxide, to destroy organic

molecules. Numerous techniques, such as Iron II catalysis, ultra-violet light, or ionizing radiation, have been used to improve oxidation by O_2.

Precipitation Processes that use techniques, such as pH adjustment, addition of carbonates or sulfides, and reducing agents, to transform a soluble compound of a metal into a less soluble form. Precipitation is used for the treatment of aqueous materials contaminated with toxic inorganic elements and compounds. Its use in the treatment of soils would normally be considered stabilization, a technology which is covered in another monograph. The procedure has been routinely used to treat wastewaters. Its application to remediation situations is less common, but data from wastewater applications are applicable.

Recently, little new activity has taken place involving precipitation processes. No specific uses of the technology were found that fit the definition of "innovative technology" used in this monograph. Therefore, these processes are not discussed herein and the reader is referred to the first monograph of this series, *Chemical Treatment* (Weitzman et al. 1994), for further information on the subject.

No information could be obtained regarding the use of substitution processes in pilot- or full-scale systems beyond the projects described in the earlier monograph (Weitzman et al. 1994). Hearsay reports of the use of the Base Catalyzed Decomposition (BCD) process for treating condensate from thermal desorption systems (Beeman 1995; Lyons 1995) were encountered. However, repeated efforts to obtain written reports or data from these field programs were unsuccessful. A development program for the BCD process is currently underway at the Naval Facilities Engineering Service Center in Port Hueneme, California. The process under development consists of a rotary reactor operating at 343°C (650°F) and a chemical treatment unit. According to the developer, the soil, mixed with 5 to 10% sodium bicarbonate is fed to the rotary reactor. PCBs are driven out of the soil and collected by condensation into a stirred tank reactor where they are chemically destroyed. According to the information submitted by IT Corporation, the contractor performing the work, efforts to date have focused primarily on the rotary reactor portion of the process. For further information, the reader is referred to the: Naval Facilities Engineering Service Center, 1100 23rd Avenue, 414ST, Port Hueneme, CA 93043.

Because the applicability of substitution processes is limited to special or isolated cases combined with the lack of field data, substitution processes are not addressed in this monograph.

While this monograph focuses on innovative treatment methods, many traditional oxidation and other wastewater treatment technologies are also applicable to contaminated site remediation. Commonly used chemical oxidants are air (oxygen), chlorine compounds (hypochlorous acid, chloramines, chlorine dioxide, bromine chloride), permanganate, ozone, hydrogen peroxide, and Fenton's reagent. Principles and applications of chemical oxidation can be readily found in the literature (Weber 1972; Glaze 1990; Cornwell 1990; James M. Montgomery Consulting Engineers 1985). Chlorine compounds have been frequently used for disinfection in water treatment. Other contaminants and undesirable properties that are amenable to chemical oxidation are iron, manganese, cyanide, phenols, taste and odor, color, disinfection byproducts, and other synthetic organics. White (1972) described the use of chlorine for treating potable water, wastewater, and cooling water. Design considerations for iron and manganese removal and taste and odor control are described in *Water Treatment Plant Design* (American Society of Civil Engineers and American Water Works Association 1990).

Coagulation/flocculation has been used to coalesce small colloidal particles (clay and silt particles in natural water, chemical precipitates, etc.) to form larger aggregates that can be removed by sedimentation and filtration. The stability of colloidal particles is controlled by electrostatic interactions and has been described using the theory of *electrical double layer* which is directly related to the phenomenon associated with electroosmosis covered in this monograph. Principles and applications of coagulation/flocculation are described in terms of coagulant types (inorganic and organic), destabilization mechanisms, and design considerations (Weber 1972; Amirtharajah and O'Melia 1990; James M. Montgomery Consulting Engineers 1985; American Society of Civil Engineers 1992; American Society of Civil Engineers and American Water Works Association 1990).

Chemical precipitation precedes coagulation in removing dissolved metal contaminants such as iron, manganese, hardness (calcium and magnesium), phosphorus, and various heavy metals. Principles and applications of this process can be readily found in the literature (Benefield and Morgan 1990; James M. Montgomery Consulting Engineers 1985; Snoeyink and Jenkins

1980; American Society of Civil Engineers and American Water Works Association 1990). The removal of phosphorus from municipal wastewater is described in *Design of Municipal Wastewater Treatment Plants* (American Society of Civil Engineers 1992).

Ion exchange has been used to remove inorganic ions (ammonium, heavy metal, etc.). Ion-exchange materials are either inorganic (e.g., zeolite) or organic (organic-polymer-based synthetic resins). Principles and design factors of ion exchange can be readily found in the literature (Weber 1972; Clifford 1990; Helfferich 1962; James M. Montgomery Consulting Engineers 1985).

Adsorption has been used to remove dissolved organics from water. The most commonly used adsorbent is activated carbon, while some synthetic resins are also used for selective organic compounds. The use of granular and powdered activated carbon and synthetic resins for removing organics was described by Weber (1972), Snoeyink (1990), and in James M. Montgomery Consulting Engineers (1985). Design considerations and equipment selections were described in *Design of Municipal Wastewater Treatment Plants* (American Society of Civil Engineers 1992) for municipal wastewater and in *Water Treatment Plant Design* (American Society of Civil Engineers and American Water Works Association 1990) and *Water Treatment Principles and Design* (James M. Montgomery Consulting Engineers 1985) for potable water.

1.2 Chemical Treatment

The term chemical treatment, as used in this monograph, refers to the use of reagents or electricity to destroy or chemically modify target contaminants by means other than pyrolysis, combustion, wet-air oxidation, solidification, or stabilization. Those are specialized forms of chemical treatment discussed in other monographs within the WASTECH® Series.

For the purpose of this monograph, chemical treatment is defined as having the following goals:

a. convert the hazardous constituents into a less toxic or environmentally less-objectionable form.

b. convert the hazardous constituents into a less mobile form, for example, by precipitation;

c. convert the hazardous constituent into a more mobile form, improving the performance of a second treatment process that removes the modified hazardous constituent from the nonhazardous matrix (While such mobilization is conceptually possible, no commercial or developmental chemical processes for doing this were identified, and the concept is not covered herein.);

d. convert the hazardous constituent into a form that is more amenable to subsequent treatment by another process. An example is the partial oxidation of contaminants in groundwater to convert refractory (difficult-to-degrade) organics into compounds that are amenable to biodegradation.

For organic contaminants, the ideal goal is complete mineralization — for example, conversion of PCBs to sodium chloride, carbon dioxide, and water. Realistically, however, the goal of most chemical treatment processes is more modest; it is the conversion of selected target contaminants into unregulated or less toxic chemical forms. For example, replacing chlorine on a PCB or chlorodibenzodioxin (dioxin) molecule with an aryl or alkyl group using, for example, a sodium naphthalide reagent or with another functional group such as a polyethylene glycol. Placement of the chlorine legally converts the hazardous compound to a nonregulated substance. In many cases, the long-term stability or environmental effects of such treatment is not well-understood. For example, Hong et al. (1995) studied the genotoxicity profiles of treated extracts from the dehalogenation of wood preserving waste using the KPEG process (see Weitzman et al. 1994 for a description of the KPEG process). Results showed that the KPEG process effectively dehalogenated the pentachlorophenol in the wood preserving waste and that the genotoxicity of the waste was reduced throughout the dechlorination reaction. However, the genotoxicity was not completely eliminated and further treatment was recommended to completely detoxify the waste.

For inorganic contaminants, the ideal goal is conversion to a nonhazardous form. This can be achieved with elements such as chromium that have a highly hazardous (i.e., hexavalent chromium) and a relatively nonhazardous (i.e., trivalent chromium) oxidation state or with organometallic compounds such as nickel carbonyl. When complete conversion is not possible, chemical treatment operates to convert metals to a less soluble, and hence, a less leachable form. This latter goal impinges on stabilization and solidification, treatment technologies that are covered in another monograph in this Series by that name.

Chemical treatment is rarely used as the sole process. It may be used as a pretreatment technique to enhance the efficiency of subsequent processes or as a posttreatment step to polish an effluent. For example:

- various advanced oxidation techniques have been successfully employed to soften organic compounds to improve their biodegradability;
- chemical dechlorination can be used to treat the contaminated eluate from solvent extraction of chlorinated organics from soil; and
- chemical destruction can be used to treat the offgas from a vapor-phase extraction process.

Chemical treatment must be performed with a knowledge of the chemical reactions involved. Also, the nature of the treated material is an important consideration when using chemical treatment. When the reagents are mixed with the contaminated material to destroy or modify the target contaminants, the "decontaminated material" still contains the chemical reaction products and any residual reagent. These remains, which are usually mobile, may be toxic, have a significant environmental impact on the surrounding ecosystem, or pose legal or safety concerns. These impacts of chemical treatment remains have been a major impediment to its use for direct treatment of soils.

Chemical treatment is technique — rather than process — oriented. To determine the proper treatment method it is first necessary to identify the target contaminant and then determine its availability and reactivity. Chemical knowledge is used to ascertain the type of chemical reactions to which the target compound(s) is amenable, to evaluate the available equipment, and to select or design the appropriate treatment system — this monograph is organized in a similar manner, grouping technologies by types of chemical reactions used.

The technologies discussed in this monograph are shown in Table 1.1. They have been grouped based on whether their application is in situ or ex-situ. All of the techniques are based on some form of electron transfer, where the target compound is (1) made available for treatment, (2) converted to a nonregulated form (this process should not be encouraged unless those responsible will do the work necessary to be certain that the nonregulated form is in fact nonhazardous to public health and the environment), or (3) oxidized completely. Electrical processes are used to migrate materials to a collection point, remove chlorine atoms from organic compounds, and coagulate and oxidize the target containments. The other processes for which detailed design and application data are provided are supercritical water oxidation and high voltage electron beam treatment, both ex-situ processes.

Appendix A discusses ex-situ chemical treatment processes which are classified as emerging technologies. These technologies have been extensively studied, but insufficient commercial application data exists to fully discuss them in the detail required. They are electrochemical coagulation (alternating-current electrocoagulation processes), and electrochemical oxidation/reduction (the Silver (II) process).

1.3 Development of the Monograph

1.3.1 Background

Acting upon its commitment to develop innovative treatment technologies for the remediation of hazardous waste sites and contaminated soils and groundwater, the U.S. Environmental Protection Agency (US EPA) established the Technology Innovation Office (TIO) in the Office of Solid Waste and Emergency Response in March, 1990. The mission assigned TIO was to foster greater use of innovative technologies.

In October of that same year, TIO, in conjunction with the National Advisory Council on Environmental Policy and Technology (NACEPT), convened a workshop for representatives of consulting engineering firms, professional societies, research organizations, and state agencies involved in remediation. The workshop focused on defining the barriers that were impeding the application of innovative technologies in site remediation projects. One of the major impediments identified was the lack of reliable data on the performance, design parameters, and costs of innovative processes.

The need for reliable information led TIO to approach the American Academy of Environmental Engineers®. The Academy is a long-standing, multidisciplinary environmental engineering professional society with wide-ranging affiliations with the remediation and waste treatment professional communities. By June 1991, an agreement in principle (later formalized as a Cooperative Agreement) was reached providing for the Academy to manage a project to develop monographs describing the state of available innovative remediation technologies. Financial support was provided by the US EPA, U.S. Department of Defense (DoD), U.S. Department of Energy (DOE), and the Academy. The goal of both TIO and the Academy was to develop monographs providing reliable data that would be broadly recognized and accepted by the professional community, thereby eliminating or at least minimizing this impediment to the use of innovative technologies.

The Academy's strategy for achieving the goal was founded on a multiorganization effort, WASTECH® (pronounced Waste Tech), which joined in partnership the Air and Waste Management Association, the American Institute of Chemical Engineers, the American Society of Civil Engineers, the American Society of Mechanical Engineers, the Hazardous Waste Action Coalition, the Society for Industrial Microbiology, and the Water Environment Federation, together with the Academy, US EPA, DoD, and DOE. A Steering Committee composed of highly-respected representatives of these organizations having expertise in remediation technology formulated the specific project objectives and process for developing the monographs (see page iv for a listing of Steering Committee members).

By the end of 1991, the Steering Committee had organized the Project. Preparation of the initial monographs began in earnest in January, 1992, and the original eight monographs were published during the period of November, 1993 through April, 1995. In Fall of 1994, based upon the receptivity of the industry and others of the original monographs, it was determined that a companion set, emphasizing the design and applications of the technologies, should be prepared as well. At this time the Soils Science Society of America joined the WASTECH® conservation. Task Groups were identified during 1995 and work commenced on this second series.

Table 1.1
Technologies Reviewed

Process	Contaminant Types	Process Description	Means of Treatment
1. In Situ Electromigration, Electroosmosis	All dissolved organics and inorganics	Electrodes are embedded in a contaminated site and a DC voltage is applied between them. In the simplest applications, only two electrodes, an anode and a cathode, are used; larger sites require anodes and multiple cathodes.	Metals and soluble organic chemicals migrate to the electrodes where they concentrate in the groundwater which is pumped out to treatment.
2. In Situ Electrochemical Reduction	Chlorinated organic compounds and oxidized metals	A permeable barrier consisting of iron metal powder is installed downgradient of the contaminated site. Impermeable barriers may be added to funnel the groundwater through the permeable barrier.	The iron in the barrier reacts with the chlorinated organic compounds in the water to remove the chlorine.
3. Ex-Situ Oxidation by Supercritical Water Oxidation	All organic compounds and inorganic salts	An aqueous stream containing (usually) high concentrations of organic materials is mixed with an oxidant (usually oxygen gas or hydrogen peroxide) in water at temperatures in the range of 350°C (662°F) to 600°C (1,112°F) and pressures of 17 MPa (2,500 psi, 170 atm) or greater.	The organics are oxidized; inorganic salts precipitate out at these conditions.
4. Ex-Situ Destruction by Electron Beam Irradiation	All organic compounds and some inorganic ions	An aqueous stream containing relatively low concentrations of organic materials is passed over a weir at ambient temperature and pressure. The cascading film of water is irradiated with a scanning high-energy electron beam (analogous to a cathode ray TV tube). The electron beam forms hydrogen and hydroxyl radicals which react with the organic compounds. Metals in solution may have their oxidation state's altered.	The organic compounds are mineralized to a high level. Metal oxides or hydroxides may precipitate out.

1.3.2 Process

For each of the series, the Steering Committee decided upon the technologies, or technological areas, to be covered by each monograph, the monographs' general scope, and the process for their development and appointed a task group of experts to write a manuscript for each monograph. The task groups were appointed with a view to balancing the interests of the groups principally concerned with the application of innovative site and waste remediation technologies — industry, consulting engineers, research, academe, and government .

The Steering Committee called upon the task groups to examine and analyze all pertinent information available, within the Project's financial and time constraints. This included, but was not limited to, the comprehensive data on remediation technologies compiled by US EPA, the store of information possessed by the task groups' members, that of other experts willing to voluntarily contribute their knowledge, and information supplied by process vendors.

To develop broad, consensus-based monographs, the Steering Committee prescribed a twofold peer review of the first drafts. One review was conducted by the Steering Committee itself, employing panels consisting of two members of the Committee supplemented by at least four other experts (see *Reviewers,* page iii, for the panel that reviewed this monograph). Simultaneous with the Steering Committee's review, each of the professional and technical organizations represented in the Project reviewed those monographs addressing technologies in which it has substantial interest and competence.

Comments resulting from both reviews were considered by the Task Group, appropriate adjustments were made, and a second draft published. The second draft was accepted by the Steering Committee and participating organizations. The statements of the organizations that formally reviewed this monograph are presented under *Reviewing Organizations* on page v.

1.4 Purpose

The purpose of this monograph is to further the use of innovative chemical treatment site remediation and waste processing technologies, that is, technologies not commonly applied, where their use can provide better, more cost-effective performance than conventional methods. To this end, the monograph documents the current state of chemical treatment technology.

1.5 Objectives

The monograph's principal objective is to furnish guidance for experienced, practicing professionals, and users' project managers. The monograph, and its companion monograph, are intended, therefore, not to be prescriptive, but supportive. It is intended to aid experienced professionals in applying their judgment in deciding whether and how to apply the technologies addressed under the particular circumstances confronted.

In addition, the monograph is intended to inform regulatory agency personnel and the public about the conditions under which the processes it addresses are potentially applicable.

1.6 Scope

The monograph addresses innovative chemical treatment technologies that have been sufficiently developed so that they can be used in full-scale applications. It addresses all aspects of the technologies for which sufficient data were available to the Chemical Treatment Task Group to review the technologies and discuss their design and applications. Actual case studies were reviewed and included, as appropriate.

The monograph's primary focus is site remediation and waste treatment. To the extent the information provided can also be applied elsewhere, it will provide the profession and users this additional benefit.

Application of site remediation and waste treatment technology is site specific and involves consideration of a number of matters besides alternative technologies. Among them are the following that are addressed only to the extent that they are essential to understand the applications and limitations of the technologies described:

- site investigations and assessments;
- planning, management, specifications, and procurement;
- regulatory requirements; and
- community acceptance of the technology.

1.7 Limitations

The information presented in this monograph has been prepared in accordance with generally recognized engineering principles and practices and is for general information only. This information should not be used without first securing competent advice with respect to its suitability for any general or specific application.

Readers are cautioned that the information presented is that which was generally available during the period when the monograph was prepared. Development of innovative site remediation and waste treatment technologies is ongoing. Accordingly, postpublication information may amplify, alter, or render obsolete the information herein about the processes addressed.

This monograph is not intended to be and should not be construed as a standard of any of the organizations associated with the WASTECH® Project; nor does reference in this publication to any specific method, product, process, or service constitute or imply an endorsement, recommendation, or warranty thereof.

1.8 Organization

This monograph is organized under a uniform outline and addresses the design and application of five innovative chemical treatment technologies available for site remediation.

For each, the following are discussed:

- scientific principles on which the technology is founded;
- potential applications of the technology;
- treatment trains, including definition of the point of application in a complete remediation and essential pre- and posttreatment processes;
- design related guidance covering:
    - remediation goals,
    - design basis,
    - design and equipment selection,
    - process modifications,
    - pretreatment processes,
    - posttreatment processes,
    - process instrumentation and controls,
    - safety requirements,
    - specification development,
    - cost data,
    - design validation,
    - permitting requirements, and
    - performance measures.
- implementation and operation issues including:
    - implementation strategies,
    - start-up procedures,
    - operations practices,

  - operations monitoring, and
  - Quality Assurance/Quality Control; and
- case histories of laboratory- and pilot-scale applications of the technology.

2
IN SITU ELECTROCHEMICALLY INDUCED PROCESSES

2.1 Scientific Principles

In situ electrochemical remediation uses electric current and potential to enhance the transport of contaminants in groundwater or to convert metal compounds to less mobile precipitates. To apply the technology, electrodes are embedded in the contaminated region and a DC voltage is applied between them. In the simplest applications, only two electrodes, an anode and a cathode, are used; larger sites require multiple anodes and cathodes. The electric current between the anodes and cathodes:

- electrolyzes a fraction of the groundwater, forming acidic and caustic zones near the anodes and cathodes, respectively;
- increases the relative mobility of some soluble organic and inorganic compounds, causing the compounds to migrate to the electrodes; and
- reduces the solubility of some metals near the cathodes by forming less-soluble metal hydroxides or carbonates.

Once soluble contaminants concentrate at the electrodes, the contaminated groundwater is pumped out and treated. In theory, insoluble organic compounds could migrate as oil droplets. However, the low permeability of most soils makes this impractical.

A variation of electrochemical remediation replaces the electrodes with a sacrificial grid or powdered metal that is embedded in an area where contaminated groundwater passes through and makes contact with the powdered metal (O'Hannesin and Gillham 1992; Gillham and O'Hannesin 1994). The metal is oxidized, serving as an electrochemical galvanic half-cell, while contaminants are reduced to complete the electrochemical cell. This variation is called permeable barrier treatment.

The concept of using in situ electrodes to chemically oxidize or reduce contaminants remains a possibility. Laboratory tests using ex-situ electrochemical cells have shown that it is possible to destroy organic compounds such as PCBs (Zhang 1995) and other organics (see the discussion of the Silver (II) process, Appendix A) by a variety of chemical means. No data beyond the laboratory phase were found. At this stage, the use of induced electric current in situ appears to be restricted to improving the mobility of metals and organic compounds and the following discussion is restricted to this application.

The fundamental electrochemical processes whereby contaminants are removed or destroyed are electromigration, electrophoresis, electroosmosis, electrocoagulation, and electrochemical reduction. *Electromigration* occurs when a charged ion in solution is transported under an electric field, whereas *electrophoresis* occurs when, instead of an ion, a charged particle is involved. However, since the movement of colloidal particles in compacted, low-permeability soils is not practical, electrophoresis is not usually considered in remediation. *Electroosmosis* occurs when a thin liquid layer around a charged particle, which contains charged ions, moves relative to a stationary and oppositely-charged surface under an electric field. *Electrocoagulation* occurs when metal ions, which come from the electrochemical oxidation of an anodic metal such as iron or aluminum, are used as coagulants for the coagulation of contaminant metal ions. This process has not been used for in situ remediation. *Electrochemical reduction* occurs when a sacrificial metal (e.g., iron) is oxidized to induce the chemical reduction of organic compounds (e.g., chlorinated organic solvents). Electrochemical reduction has been tested in the form of permeable barriers for treatment of solvent-contaminated groundwater. Permeable-barrier treatment that does not require an external input of electricity is covered in Chapter 3. Electromigration and electroosmosis that require an external input of electricity are covered in this chapter.

2.1.1 Electromigration

The process of electromigration has been described by many researchers and developers (Acar 1992; Acar 1993; Acar, Alshawabkeh, and Gale 1993; Acar et al. 1995; Lindgren, Mattson, and Kozak 1994; Marks, Acar, and Gale 1994; Mattson and Lindgren 1995; Probstein and Hicks 1993). In this process, an array of electrodes (cathodes and anodes) is inserted into soil with a potential difference on the order of a few hundred V/m (Acar and Alshawabkeh 1993). In this electric field, cations (e.g., metal ions) move to the cathodes whereas anions (e.g., cyanide complexes, metal-hydroxide complexes, anionic dyes, and chromate) move to the anodes (see Figure 2.1).

Figure 2.1
Schematic of Electromigration

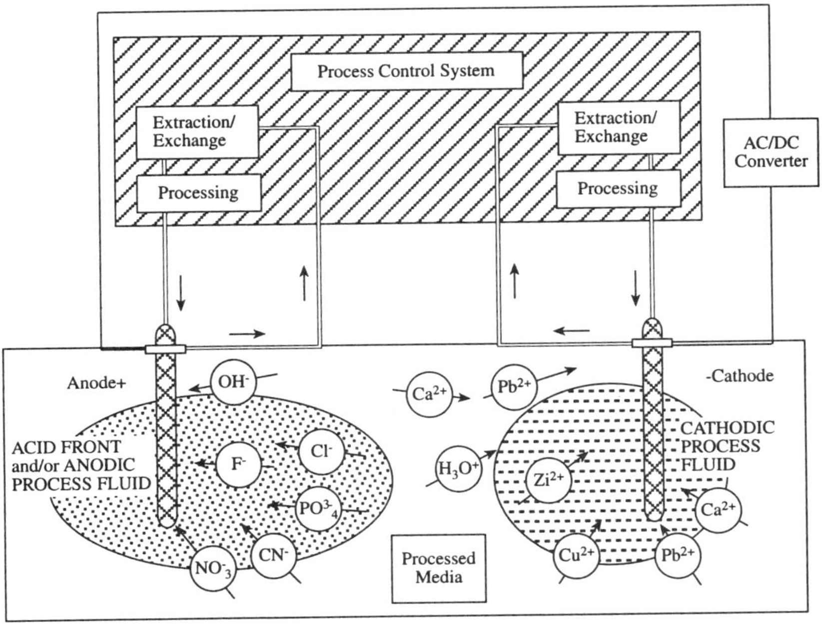

Reprinted from Journal of Hazardous Materials, Volume 40, Acar et al., "Electrokinetic Remediation: Basis and Technology Status," pp 117-137, 1995 with kind permission of Elsevier Science - NL, Sara Burgerhartstraat 25, 1055 KV Amsterdam, The Netherlands.

At the electrodes, electrolysis of water takes place as follows:

At the anode,

$$2H_2O \rightarrow 4H^+ + O_2(g) + 4e^- \tag{2.1}$$

At the cathode,

$$2H_2O + 2e^- \rightarrow 2OH^- + H_2(g) \tag{2.2}$$

This electrolysis results in an acidic front at the anode and an alkaline front at the cathode, which move to the cathode and the anode, respectively. The propagation of the acid and base fronts promotes the dissolution of metal ions near the anode and the precipitation of the metal ions near the cathode. These conditions significantly affect (1) the pH and ionic strength of pore water, (2) the mobility and solubility of metal contaminants, (3) charge conditions of soil particles, and (4) the hydraulic conductivity of the porous media. Depending on the type of contaminants of concern, these conditions could be significant or minimal. For example, the pH could drop to around 2 at the anode and increase to around 12 at the cathode (Acar and Alshawabkeh 1993; Kahn and Alam 1993).

The pH affects the precipitation equilibria of metal hydroxides and carbonates (high pH effects more precipitation). Khan and Alam (1993) demonstrated in laboratory studies the dissolution of metal precipitates (lead, manganese, and zinc) in soil due to the acid front from the anode and the subsequent migration of the dissolved metal ions to the cathode. Runnells and Wahli (1993) observed the transport of copper and sulfate ions toward the cathode and anode, respectively, and the precipitation of copper hydroxide near the cathode in a column of fine quartz. Probstein and Hicks (1993) determined that zinc removal from clay occurred mainly by electromigration and diffusion (electroosmotic flow was negligible) and that zinc precipitation occurred at the isoelectric point where the acid and base fronts converge.

Marks et al. (1994) discussed the role of H^+ in the cation-exchange equilibria of a soil-metal ion system. Hydrogen ions in the acid front displace metal ions that are adsorbed on soil particles by simple ion exchange and surface complexation mechanisms, resulting in more mobile metal ions in pore water that could be transported by electromigration. Cation-exchange capacities of common clay minerals are shown in Table 2.1. Because reduction reactions occur at the cathode, some metal ions could be reduced to elemental metals and deposited on the surface of the electrode. Therefore, in electromigration, several processes can take place simultaneously: dissolution, ion exchange, migration, precipitation, and reductive deposition.

The transport of charged metal ions under an electrical field was described as similar to diffusive transport due to a concentration gradient (Shapiro and Probstein 1993; Acar and Alshawabkeh 1993). Acar et al. (1995) reported an ionic migration of 1 to 80 cm/day (0.39 to 31.5 in./day) under an electrical field of 100 V/m.

Table 2.1
Cation-Exchange Capacities of Common Clay Minerals

Mineral	Structural Control	Exchange Capacity (meq/100 g at pH 7)
Kaolinite	Unsatisfied valences on edges of structures	3-15
Halloysite (2 H_2O)	Unsatisfied valences on edges of structures	5-15
Halloysite (4 H_2O)	Unsatisfied valences on edges of structures	40-50
Illite	Octahedral/tetrahedral substitutions, edges and K^+ deficiency between layers	10-40
Allophane	Amorphous structure, unsatisfied valences	70
Montmorillonite	Octahedral/tetrahedral substitutions and edges	70-100
Vermiculite	Replacement interlayer cations, substitutions	100-150

Source: Marks, Acar, and Gale 1994; Garrels and Christ 1965

2.1.2 Electroosmosis

The movement of a thin, charged layer is responsible for electroosmosis in a system similar to the one shown in Figure 2.1 for electromigration. The concept of an electrical double layer has frequently been used to describe electrostatic interactions between negatively charged particles (e.g., clay and silt) and positively charged ions in water (added as a coagulant) when removing the particles from water via coagulation (Amirtharajah and O'Melia 1990). The particles are negatively charged in natural water because of their negative zeta potential, -20 to -40 mV according to Amirtharajah and O'Melia (1990) and -10 to -100 mV according to Probstein and Hicks (1993).

In this double-layer concept, there are two electrical layers around a charged particle: (1) a compact layer of negative ions on the surface of the particle and (2) a diffuse layer of positive ions that are electrostatically attracted to the compact layer and somewhat dispersed into the water because of their thermal motion. This positively-charged diffuse layer (1 to 10 nm according to Probstein and Hicks (1993)) becomes mobile in electroosmosis as the layer is pulled to the cathode under an electrical field (see Figure 2.2). Electroosmosis is only effective for low-permeability, fine-grained soils that have a hydraulic conductivity of less than $1 \cdot 10^{-5}$ cm/sec ($3.9 \cdot 10^{-6}$ in./sec)(Segall and Bruell 1992). For a hydraulic conductivity greater than $1 \cdot 10^{-5}$ cm/sec, the electroosmotic effect is nullified by backflow from the cathode. The effectiveness of electroosmosis is also reduced by the production of H^+ at the anode and other sources of cations, since the diffuse layer becomes compressed as the concentration of positive ions in water increases.

Figure 2.2
Schematic of Electroosmosis

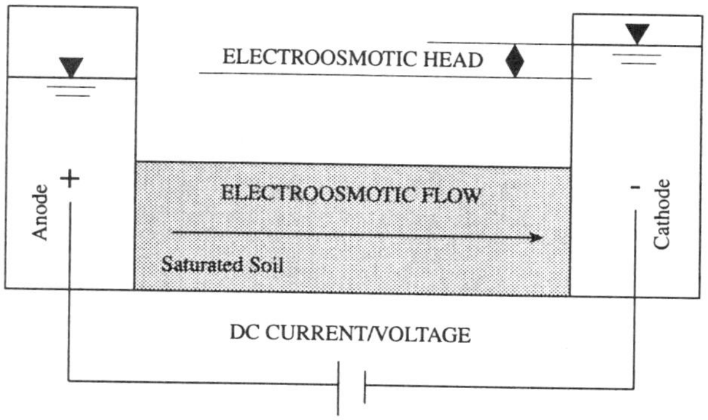

Reprinted from Journal of Hazardous Materials, Volume 40, Acar et al., "Electrokinetic Remediation: Basis and Technology Status," pp 117-137, 1995 with kind permission of Elsevier Science - NL, Sara Burgerhartstraat 25, 1055 KV Amsterdam, The Netherlands.

Electroosmosis causes a convective movement of pore water, which is different from electromigration or diffusion. Therefore, any contaminants (either ionic or neutral) in the fluid can be transported to the cathode along with the fluid. The movement of ionic species would be enhanced if they were cationic or hampered if they were anionic because of the effect of electromigration. Comparing phenol and acetic acid, Shapiro, Renauld, and Probstein (1989) attributed the lower electroosmotic flow rate for acetic acid to its lower pH, which compresses the diffuse layer. They also observed a lower removal of acetic acid when the pH was high at the cathode because of the back-migration of ionized acetic acid molecules to the anode.

The aqueous solubility and adsorbability of organic compounds can affect their removal by electroosmosis. Bruell et al. (1992) observed that organic solvents with relatively high aqueous solubility and low adsorbability (e.g., benzene, toluene, trichloroethylene, and m-xylene) were easily removed from water-saturated kaolin clay. They also observed that solvents with relatively low aqueous solubility and high adsorbability (hexane and isooctane) were not transported easily. The possibility of solubilizing relatively insoluble compounds with surfactant was suggested (Probstein and Hicks 1993). Marks, Acar, and Gale (1994) proposed an in situ bioremediation method in conjunction with electroosmosis to deliver a nutrient-containing solution by electroosmosis to microorganisms in low-permeability soil.

Electroosmotic flow is described by an equation similar to Darcy's law (Marks, Acar, and Gale 1994; Segall and Bruell 1992):

$$u_e = \frac{Q_e}{A} = k_e i_e \tag{2.3}$$

where:
u_e = electroosmotic velocity (m/sec);
Q_e = electroosmotic flow rate (m^3/sec);
A = cross-sectional area (m^2);
k_e = electroosmotic conductivity or electroosmotic coefficient of permeability (m^2/V-sec); and
i_e = electrical gradient (V/m).

The variable k_e can be related to zeta potential and the viscosity of water according to the Helmholts-Smoluchowski theory (Hunter 1982; Probstein 1989; Acar and Alshawabkeh 1993). According to Segall and Bruell (1992), k_e varies from 10^{-9} to 10^{-10} m^2/V-sec for a wide range of soils, resulting in an electroosmotic velocity of 10^{-5} to 10^{-6} cm/sec under an electric gradient of 100 V/m. The maximum electroosmotic flux was reported to be approximately 10^{-4} cm/sec at 100 V/m (Acar et al. 1995), indicating that the electroosmotic velocity is expected to be approximately 0.1 to 10 cm/day (0.04 to 4 in./day) at 100 V/m. Overall, transport of a compound is determined by the combination of electroosmosis, electromigration (if ionized), and diffusion.

The transport of metal ions in an electric field occurs because of both electroosmosis and electromigration. Although the transport of metal ions is dominated by electromigration in most cases, it is difficult to determine precisely which is the primary one because it depends on soil characteristics, applied energy level, and production and transport of hydrogen and hydroxide ions at the electrodes. Therefore, the transport of metal ions is discussed in the potential applications, Section 2.2, relative to both electroosmosis and electromigration.

2.2 Potential Applications

Electromigration involves the movement of ionic contaminants in pore water toward oppositely-charged electrodes and does not require the movement of the water being treated. Therefore, electromigration is not dependent on pore size. Although it can be applied to both high- and low-permeability soils, electromigration may not be suitable in high-permeability soils because a simple pump-and-treat method may be more convenient, flexible, and cost-effective. Thus, electromigration is typically used in conjunction with electroosmosis for low-permeability soils. Electromigration can be applied to only ionic contaminants (e.g., metal ions and dissociated organic acids and bases) and is not suitable for the removal of neutral contaminants (e.g., undissociated organic acids and bases and organic solvents).

Electroosmosis, on the other hand, does involve the movement of pore water — any contaminants that are dissolved in the water are transported to the cathode. Therefore, this process can be used for both ionic and nonionic contaminants. At the same time, electromigration will take place because of the electric field and will affect the transport of ionic contaminants.

When compared to other in situ technologies that only target one group of contaminants, organic or inorganic, electroosmosis is advantageous because it can be applied to a broad range of contaminants, forms, and concentrations. In addition, most in situ remediation technologies are ineffective for removal of contaminants from low-permeability soils (e.g., fine-grained soils). However, with electroosmosis, the electric field provides a high degree of hydraulic control. The direction of flow can be controlled by appropriately placing anodes and cathodes, and the electroosmotic flow can be initiated or stopped by applying or discontinuing electrical current. Since electroosmosis is not a pressure-driven process, channeling is also minimized.

Electroosmosis depends on porosity and zeta potential and is not affected by pore-size distribution. Acar and Alshawabkeh (1993) indicated that the maximum electroosmotic flow often occurs in low-activity clays ("activity" is defined as the plasticity index divided by the percent of clay particles less than 2 μm in size) with high water content and low ionic strength. Electroosmosis was first used in the 1930s in Germany to dewater and stabilize soils (Probstein and Hicks 1993). Stabilization occurs because consolidation through dewatering alters the physical and chemical properties of the soils (Cabrera-Guzman et al. 1990).

In summary, electroosmosis can be applied in conjunction with electromigration to remove metals and organics in low-permeability soils. Required energy input depends on soil types and conditions and types of contaminants. An acid or a base may be introduced at the electrodes as a pretreatment or as part of the overall treatment to enhance the transport and removal of contaminants.

2.3 Treatment Trains

In situ electrochemical treatment is rarely the only form of treatment at a contaminated site. This method concentrates contaminants and is usually part of a remediation program that includes one or more of the following components:

- diversion systems, such as reduced-permeability walls, to reduce groundwater infiltration to the contaminated area;
- covers or caps that reduce rainwater infiltration to the area;
- monitoring wells to allow sampling of the groundwater;

- wells enabling the injection of solutions to modify contaminant migration or the reinjection of treated water;
- wells through which groundwater is pumped to the surface to depress the groundwater level and/or remove contaminated groundwater;
- treatment systems to remove contaminants from the extracted groundwater; and
- air pollution control equipment to capture volatile organic compounds (VOCs) that may be released from the wells, water treatment system, or pumps.

2.4 Remediation Goals

The goal of electrochemically-induced processes is to concentrate the contaminants preferentially in the vicinity of the electrodes so that the concentrations of the contaminants in the contaminated zone would be below target contaminant levels.

2.5 Design

2.5.1 Design Basis

Design of an electrochemical treatment system is highly site-specific. The system's geometry is governed by the site characteristics and the properties of the soil and the contaminants. Materials of construction for the system, especially those of the electrodes, depend on the nature of the contaminants, the natural materials found at the site, and the products of electrolysis of these substances.

Once the area of contamination is determined, the contaminated soil must be defined with respect to permeability, moisture content, cation-exchange capacity, organic content, and pore water characteristics (pH, alkalinity, ionic strength, etc.). This process is most suitable for low-permeability soils. For high-permeability soils, conventional pump-and-treat methods may be more suitable and should be investigated first. An adjacent high-permeability region could adversely affect the decontamination of a low-permeability region by electroosmosis by providing a return flow (Segall and Bruell 1992). Therefore, spatial variation in permeability and contamination needs to be carefully assessed before electroosmosis is applied.

Moisture content is important because both electroosmosis and electromigration require moisture. Electromigration was reported to be effective in soil with a moisture content as low as 7% (Lindgren, Kozak, and Mattson 1991). Electroosmosis was originally used to dewater and stabilize soils, mine tailings, and mineral sediments (Probstein and Hicks 1993).

Cation-exchange capacity is important for transport of ions by electromigration and movement of the acid from the anode. Typical cation-exchange capacities of clay minerals are provided in Table 2.1. Soil with a high cation-exchange capacity is expected to slow the transport of contaminant cations to the cathode by exhibiting a high affinity for the ions and keeping the pH of the pore water from decreasing (i.e., keeping metal hydroxides and carbonates from dissolving).

The organic content of soil has been reported to be responsible for adsorbing hydrophobic organic compounds (Mills et al. 1985). The partitioning between the pore water and the soil has been frequently described by using octanol/water partition coefficients. Therefore, the transport of organics via electroosmosis will depend on the affinity of the soil for the compound of interest, especially when a purge solution is introduced to push the contaminant through the soil pore. As mentioned earlier, Bruell, Segall, and Walsh (1992) established that organic solvents with relatively high aqueous solubility and low adsorbability (relatively hydrophilic) were easier to remove than solvents with low solubility and high adsorbability (relatively hydrophobic). The use of surfactants in the purge solution may enhance the mobility of hydrophobic compounds.

The pH of the pore water affects the solubility of metal hydroxides and carbonates and thus the transport of metal ions by electromigration. The alkalinity of pore water represents its buffering capacity — high alkalinity pore water resists pH changes that could be caused by the arrival of the acid and base front from the electrodes. High alkalinity may result in:

- relatively small amounts of metal-ion transport due to low dissolution of metal hydroxides and carbonates near the anode; and
- reduced metal precipitation near the cathode.

The ionic strength of the pore water may affect the thickness of the diffuse double layer by changing the zeta potential and, therefore, the rate of electroosmotic flow. High ionic strength could be caused by either background ions or contaminant ions. Probstein and Hicks (1993) suggested that a two-step process may occur for the case of high ionic strength due to contaminant ions consisting of:

- the removal of contaminant ions by electromigration, which eventually increases zeta potential and thus the thickness of the double layer; and
- the removal of neutral contaminants (e.g., undissociated or undissociable organics) by electroosmosis.

The effect of ionic strength (especially various cations present in pore water) on the amount of metal ions adsorbed on soil may be significant.

Some ions (e.g., carbonates, phosphates, chloride, sulfides, and ammonium), which could be present in pore water or introduced as part of a purge solution, can enhance the transport of certain metal contaminants by forming soluble complexes or retard the transport by forming precipitates. It is also possible for some metal ions to become negatively-charged metal complexes and transported to the anode.

Metal ions and organic ions can be removed by both electromigration and electroosmosis, whereas neutral contaminants such as organic solvents can only be removed by electroosmosis.

2.5.2 Design and Equipment Selection

When designing an electrochemical treatment system, key considerations include electrode composition and configuration, power requirements, and purging solution selection. The electrodes should be made of materials that do not degrade under corrosive conditions or upon application of electrical current. Examples of iridium-coated titanium, zinc-coated iron wire, and galvanized steel electrodes used by Electrokinetics, Inc., are shown in Figures 2.3, 2.4, and 2.5.

Figure 2.3
Iridium-Coated Titanium Anode Used by Electrokinetics, Inc.

Reproduced courtesy of Electrokinetics, Inc. (1995)

The configuration of electrode systems with respect to the number of electrodes, spacing, and orientation needs to be considered for maximum hydraulic control. Since most contaminants to be removed are metal ions and nonionic organics, the contaminants are transported toward the cathode. In this case, an electrode system, consisting of a cathode and multiple anodes around the cathode is frequently used to collect the contaminants at the cathode (e.g., a polygon shape with a cathode at the center and anodes at the corners). The electrodes could be placed horizontally or vertically. Geokinetics, a European remediation contractor, used an electrode system that consisted of one long horizontal cathode (0.5 m (1.65 ft) below the ground surface) and a row of vertical anodes (up to 1 m (3.3 ft) deep and 1 m (3.3 ft) apart) (Lageman 1993). Bruell, Segall and Walsh (1992) suggested 3 m (10.9 ft) electrode spacing. Probstein and Hicks (1993) suggested an electrode spacing of 2 to 10 m (6.6 to 33 ft) and an electrode depth of 2 to 20 m (6.6 to 66 ft). For a relatively large area, several electrode systems can be installed.

The net flux of contaminants toward either the cathode or the anode due to electromigration, electroosmosis, and diffusion can be estimated using the equations in Section 2.1. For example, Equation 2.3 represents the contribution of electroosmosis by considering the magnitude of applied electric potential, zeta potential, viscosity of water, and electroosmotic permeability as described by Acar and Alshawabkeh (1993) and Shapiro and Probstein (1993). An ionic migration of 1 to 80 cm/day (0.4 to 31 in./day) and an electroosmotic velocity of 0.1 to 10 cm/day (0.04 to 4 in./day) at an electric gradient of 100 V/m (2.5 V/in.) were reported.

Figure 2.4
Zinc-Coated Wire Cathod Used by Electrokinetics, Inc.

Reproduced courtesy of Electrokinetics, Inc. (1995)

Power requirements can be estimated based on a desirable contaminant flux. Probstein and Hicks (1993) suggested the following ranges of power requirements:

- 40 to 200 V for applied electrical potential;
- 20 to 200 V/m (0.5 to 5 V/in.) for applied electric field strength; and
- 50 to 500 mA/cm^2 (320 to 3,200 mA/in.^{2}) for current density.

The range of current density is consistent with the values used by Electrokinetics (US EPA 1995c) and Geokinetics (Lageman 1993). Acar et al. (1995) noted that increasing current density does not necessarily increase removal efficiency.

Figure 2.5

Galvanized Steel Electrode Used by Electrokinetics, Inc.

Reproduced courtesy of Electrokinetics, Inc. (1995)

Photovoltaics may be the ideal DC source because they produce a DC current, the voltages from individual photovoltaic panels are in the appropriate range, and storage requirements are likely unnecessary (Probstein and Hicks 1993). When AC is used as a power source, the AC has to be converted to DC.

When anionic contaminants (e.g., cyanide complexes, metal-hydroxide complexes, anionic dyes, and chromate) are to be removed by electromigration, the flux toward the anode due to electromigration should be greater than the flux toward the cathode due to the combined effect of electroosmosis and diffusion.

In electroosmosis, the flow rate will eventually diminish if a purging solution is not introduced at the anode or if precipitation at the cathode decreases the efficiency of the process. In most cases, removal of one pore volume is not sufficient to remove contaminants to regulatory levels. Also, the composition of the purging solution must not adversely affect the zeta potential of the soil (and thus the thickness of electrical diffuse double layer) so that the electroosmotic flow rate can be maintained.

2.5.3 Process Modification

The main way to adapt electrochemical processes to accommodate varying site conditions is through the use of purge solutions. Purge solutions can accelerate the removal of contaminants by increasing their solubility in water. When the alkalinity of the pore water is relatively high, an acid solution can be introduced at the anode to increase the solubility of metal hydroxides and carbonates by lowering the pH of the water to an optimum value. The added acid also affects the ion-exchange equilibrium and facilitates the desorption of metal ions from soil particles. However, the strength of the acid solution must be carefully selected to preclude the resulting pH of the pore water from adversely affecting the zeta potential of the soil, and thus the electroosmotic flow rate. Because the zeta potential of a typical soil is negative, electroosmotic flow is toward the cathode. If the pore water pH is too low, the zeta potential could be reversed, causing the electroosmotic flow to flow toward the anode.

A key condition in controlling the electroosmotic flow direction is understanding the zero point of charge (ZPC). ZPC represents a pH where the zeta potential of a particle is zero. For example, kaolinites have a ZPC of 3.3 to 4.6 depending on the clay source (Parks 1967; Shapiro and Probstein

1993). Clay has a positive zeta potential below the ZPC and a negative potential above the ZPC. If the strength of the acid solution is selected such that the resulting pH of the pore water drops below the ZPC, the electroosmotic flow direction will be reversed. If the resulting pH is close to (but not below) the ZPC, the flow direction may not be reversed; however, the flow rate may decrease significantly because of a compressed double layer. Of course, other cations can also compress the double layer and thus affect the zeta potential. Therefore, the ionic strength of the purge solution and the type of cation used (e.g., the charge of the cation) also need to be considered to determine the composition of the purge solution.

The acidic purge solution may also complement the hydrogen ions produced at the anode to neutralize the hydroxide ions produced at the cathode and prevent metal hydroxides and carbonates near the cathode. For example, an acetic acid solution (0.05 M) was used in a treatment process to enhance the removal of uranyl ion and to prevent the precipitation of the ion near the cathode (Acar and Alshawabkeh 1993) (see Figure 2.6). When the electroosmotic flow rate is low, an acid solution or water may be introduced around the cathode to flush or neutralize the hydroxide ions produced, maintain a neutral pH, and prevent the metal ions from precipitating.

Flushing the hydrogen ions produced at the anode may be necessary when the resulting pore water pH is too low to cause an adverse impact on electroosmotic flow or on the integrity of the clay mineral structure by dissolving silica and alumina. An alkaline solution or water can be used for this purpose.

Instead of using an acid or alkaline solution to control the production of hydrogen and hydroxide ions at the anode and cathode, respectively, the use of membrane or ion-exchange materials around the electrodes has been suggested (Probstein and Hicks 1993; Marks, Acar, and Gale 1994). The use of a chelating agent (e.g., ethylenediamine tetraacetic acid [EDTA]) was reported to enhance the mobility of metals by forming metal-EDTA complexes (Allen and Chen 1993). The use of a surfactant solution as a purge solution was also suggested to enhance the solubility of relatively hydrophobic organic contaminants (Marks, Acar, and Gale 1994).

The use and selection of a purge solution needs to be carefully evaluated in terms of regulatory requirements and potential environmental impacts. A selected purge solution has to be acceptable to regulatory agencies, and all precautions have to be taken to minimize potential dispersal of contaminants beyond the zone of contamination.

Figure 2.6
Precipitation of Uranium Hydroxide at Cathode
("Yellow Cake")(Electrokinetics, Inc.)

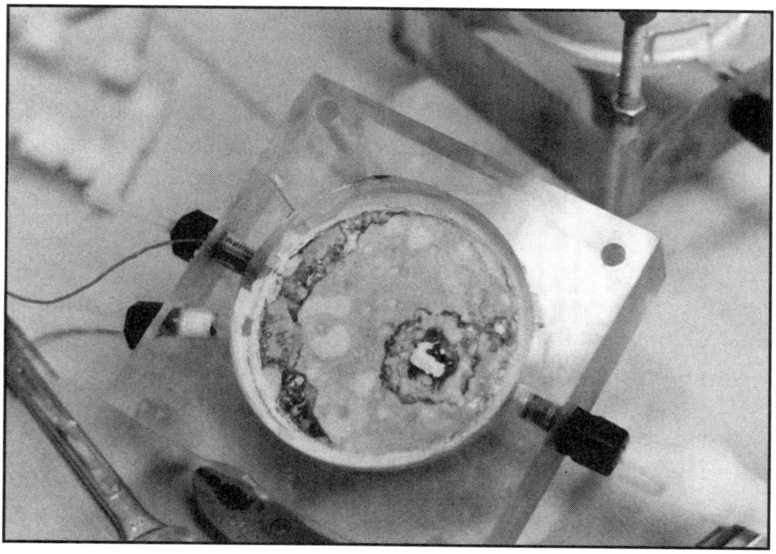

Reproduced courtesy of Electrokinetics, Inc. (1995)

2.5.4 Pretreatment Processes

Lageman (1993) suggested removal of conducting objects of larger than 10 cm (4 in.) (e.g., tins, barrels, reinforcing rods) as a pretreatment whenever possible because these objects may function as preferential flow paths for the electrical current and delay the movement of contaminants. He also indicated that nonconducting objects (e.g., wooden beams, plastic sheets, concrete blocks) may interfere with electroosmosis/electromigration. In addition, any subsurface pipes and cables should be located prior to treatment so that they can be protected.

Pretreatment methods which can improve the mobility of groundwater through the soil or to improve the contaminants' solubility and hence their mobility should be evaluated. In addition, solutions of acids or caustics can be injected into the contaminated formations to open up the soil structure. Such pH modification could also increase the solubility, and hence the mobility of metal contaminants.

2.5.5 Posttreatment Processes

In electrochemical processes, the groundwater collected from the region around the cathode (or the anode if the contaminants of concern are anionic) will be concentrated with contaminants. The groundwater must be pumped to the surface, collected, treated, and disposed. The treatment methods used will depend on the types of contaminants at the site. If the contaminants are organic, the liquid can be treated using many traditional separation or destruction processes such as adsorption, stripping, biological treatment, chemical or photochemical oxidation, or any of their combinations. If the contaminants are inorganic, they could be separated by chemical precipitation, ion exchange, or reverse osmosis. The description of these processes can be readily found in a number of books, papers, and reports listed in Appendix B.

The recovery of some purging agents (e.g., complexing agents and surfactants) may be necessary. Allen and Chen (1993) described an electrolytic process to recover EDTA from an EDTA-lead solution which was used to chemically extract lead from a contaminated soil. In the recovery process, as EDTA-lead complexes are electrolytically destroyed, the lead is deposited on a copper cathode while the EDTA is released. To prevent the EDTA from being electrolytically oxidized at the anode, the anode was separated from the solution by a cation-exchange membrane. This recovery process may be directly applied to in situ electromigration/electroosmosis by recovering heavy metals at the cathode as a deposit and returning the EDTA solution to the anode, which is separated by a cation-exchange membrane.

Used surfactants may be recovered by ultrafiltration, a membrane separation process, or by sieving out organic contaminants which are generally larger than surfactants. However, the cost of recovering spent purging agents should be compared with that of continuously adding the agents and treating them as contaminants.

2.5.6 Process Instrumentation and Control

The power applied and the composition of purging agents need to be monitored and controlled based on design parameters. Additionally, the effectiveness of the process needs to be assessed by monitoring the quality of treated groundwater (see discussion regarding design validation [Section 2.5.10]).

2.5.7 Safety Requirements

The potential health hazards associated with in situ electrochemical site remediation are generally chemical in nature and involve the contaminants of concern (both organics and inorganics) and the chemicals used for remediation. To protect a worker from the chemical hazards, the threshold limit values, the short-term exposure limits, and the immediate danger to life and health levels for the chemicals involved need to be identified. After determining these levels for the chemicals involved in all forms (vapors, dust, and liquids), appropriate control actions need to be taken to provide a safe environment for workers. Depending on the types and nature of the chemicals involved, protective devices (e.g., respirators, gloves, clothing, boots, safety glasses, etc.) must be specified to protect the workers against potential hazards to the respiratory system, skin, and eyes.

The use of electricity also presents a hazard either as a direct contact with electrified positive anodes or contact with soil surfaces that are in contact with electricity. The treatment areas where the maximum voltage is expected to be over 2 to 10 V should be fenced to exclude entry while power is applied. In addition, warning signs about the hazards associated with electricity need to be posted. Before work is performed using any equipment at the site, the electricity should be turned off to avoid any electric shock. Use of protective devices to avoid electric shock is also necessary.

2.5.8 Specification Development

The key requirements that must be incorporated in specifications for an electrochemical treatment application depend on whether the bids are for equipment which is to be installed by others, or for a turnkey electrochemical system with performance guarantees.

If a vendor is to provide a turnkey system with performance guarantees, such guarantees must be based on the overall site characterization and the overall treatment scheme (beyond just the electrochemical system) that is to be used at the site. The site description should incorporate all of the issues discussed in the design checklist in Section 2.5.13. Namely, the vendors should have a clear understanding of the nature of the contaminants and how they are distributed. It is important to provide information on the actual chemical form of the contaminants. For example, simply specifying that a site contains mercury is insufficient. Mercury can be found as a metal, as methyl mercury, or in another chemical form; each form will behave in an entirely different way under the influence of the electric field. To successfully integrate an electrochemical process into the overall remediation, the vendor must be apprised of the overall scheme. An in situ electrochemical process is only part of a system that can include infiltration controls, groundwater diversion systems, injection wells, and pumping systems.

Specification of individual equipment to be installed (for example, electrodes and power supply) is fairly straightforward. Electrical equipment should be specified to allow for a far wider variation in operation than that envisioned in the system design. For example, the DC power supply must be capable of withstanding a sudden rise in the level of the groundwater that would otherwise cause it to short circuit. The types of electrical safeguards required to prevent destruction of the power supply must be clearly defined to the vendors.

Electrode specifications must clearly state the types of metals and alloys used for different portions. Corrosion resistance is a crucial concern. It is relatively easy to reduce the cost of the electrodes by, for example, shortening the length of the parts made of a unique metal; however, this could be offset by a reduced electrode life.

It is preferable to give equipment vendors performance, rather than design, specifications. Vendors may have *proprietary systems* that can meet the performance requirements at a lower cost. Vendor involvement in the specification process is essential.

2.5.9 Cost Data

Cost data for electrochemical remediation are still being developed; however, the processes appear to be competitive with other in situ processes.

Installation and equipment costs for electrochemical systems are relatively modest. The major operating cost elements (above those for a normal groundwater flow modification and pumping system) are electric power and electrode replacement.

Probstein and Hicks (1993) estimated energy costs for electroosmosis to be approximately \$1.10/tonne (\$1/ton) of soil treated using:

- a value of 20 kWh/m^3 (15kWh/yd^3), which was taken from the range of 10 to 20 kWh/m^3 (8 to 15 kWh/yd^3) reported by Segall and Bruell (1992);
- \$0.10/kWh,
- a soil porosity of 50%;
- a dry specific gravity of 3; and
- two pore volumes of purging.

They also estimated the energy costs for electromigration to be approximately \$2.20/tonne (\$2/ton) — an estimate that is more problematic because electromigration is concentration-dependent and voltage-specific. This estimate was based on power requirements for electromigration of approximately 40 kWh/m^3 (30 kWh/yd^3) as reported by Hamed, Acar, and Gale (1991).

Considering a safety factor of 10, Probstein and Hicks (1993) predicted the energy costs for electroosmosis in conjunction with electromigration to be approximately \$22 to \$33/tonne (\$20 to \$30/ton). This is comparable with an energy cost estimate of \$16 to \$33/tonne (\$15 to \$30/ton) with a power usage of 60 to 200 kWh/m^3 (46 to 150 kWh/yd^3) for removing lead from kaolinite specimens (Acar et al. 1995). However, it is not clear why these cost estimates were so close for quite different values of power usage. The cost is expected to vary substantially because of a wide range of reported power usage: 18 to 39 kWh/m^3 (14 to 30 kWh/yd^3) for removing phenol (Acar, Li, and Gale 1992); 200 kWh/m^3 (150 kWh/yd^3) for removing hydrocarbons (Bruell, Segall, and Walsh 1992); 300 to 700 kWh/m^3 (230 to 540 kWh/yd^3)(US EPA 1995c), and 65 to 300 kWh/m^3 (50 to 230 kWh/yd^3)(Lageman 1993). Some of these values are an order of magnitude higher than the value used by Probstein and Hicks.

The cost of electrode replacement depends to a large extent on the characteristics of the water and contaminants at the site and can vary widely. No meaningful general estimate of this cost can be made at present.

2.5.10 Design Validation

In almost all applications, electrochemical remediation is part of a long-term remediation program. As such, it is crucial that the design be validated early in the program. The system's purpose is to concentrate contaminants at the electrodes. Hence, the most direct validation program is to measure the contaminant concentration in the groundwater in and around the contaminated zone before and after an electric potential is applied to the electrodes. The data analysis would validate whether or not the treatment is effective in removing the contaminants and also that the contaminants stay in the zone. Such a sampling and analysis program must use statistical techniques that account for normal variability in contaminant concentration. Several techniques can be used to validate the design.

The simplest would be to monitor the discharge wells around the electrodes for a period of one year prior to energizing them. Sampling times would be chosen to be representative of a variety of weather and climatic conditions. A one-year interval would allow for sample collection during the various seasons. At the start of the second year, the electrodes would be energized and samples would be collected at times that appear to best duplicate the site conditions during the first year of sampling. Statistical analysis of results from the two sets of samples will identify if the treatment is effective or if further adjustment is warranted.

The above validation program has several limitations. One is that there is no guarantee that the weather conditions between the years would be the same. Such variability could mask the system's performance. Another limitation is that it may take a long time to determine whether remediation system enhancement is necessary.

An alternative validation scheme is to implement a cyclic technique. Initially, samples would be collected for a few days after installing but before energizing the electrodes. Then, the system would be energized for the same number of days, the sampling would be repeated, and the system would be shut off. This process could be repeated as necessary for validation. The length of time for the cycles should be greater than the time required for contaminants to migrate to the electrode regions. This "pulsed" validation technique should reduce the impact of seasonal variability on sampling and analysis results and can be used to fine-tune the system's operation, e.g., the system could be adjusted based on the contaminants' migration velocities.

2.5.11 Permitting Requirements

The types of information that must be supplied to the regulatory agencies to justify use of electrochemical processes will generally be the same as that required for a traditional pump-and-treat system. In addition, it will be necessary to provide complete information on the design and operating conditions for the system, chemicals (chelating agents for metals, buffering agents for electrodes, etc.) to be used for treatment, and any potential byproducts such as evolution of gases at cathodes. Finally, it will be necessary to demonstrate that the system can achieve the remediation goals without adverse local environmental impact.

2.5.12 Performance Measures

The intent of the electrochemical treatment systems is to increase the flow of the contaminants to wells for removal from the site. Therefore, the main performance measure for such systems is simply the amount of contaminants removed per unit of groundwater pumped for treatment.

2.5.13 Design Checklist

A. Local conditions, soils, and geologic formations

1. Porosity and permeability of the soil(s) in the system area of influence
2. Location of natural and artificial barriers (buried metallic or non-metallic objects) to flow and electric current and their impact on the migration of groundwater and contaminants toward the electrodes
3. Changes in water levels that might short electrodes
4. Naturally-occurring materials (i.e., salt) that might impact the process
5. Special site-specific requirements to protect the system from weather

B. Contaminant types

1. Presence of inorganic materials that might clog pores over time because of electrochemical precipitation
2. Behavior of all contaminants (not just those targeted) under the influence of the electric field
3. Interaction between contaminants and naturally-occurring materials at the site

C. Electrode composition

1. Corrosion characteristics of the electrodes
2. Anticipated lifetime in situ
3. Effect of the electric current on electrode corrosion
4. Effect of buffering solution on electrodes

D. Well casing materials

1. Corrosion characteristics
2. Anticipated lifetime in situ
3. Effect of the electric current on corrosion
4. Effect of the casing on the electric field strength and shape

E. Power requirements

1. Operating voltage range and allowable variation
2. Commercial availability of suitable power supplies
3. Local availability of sufficient power

2.6 Implementation and Operation

2.6.1 Implementation Strategies

In situ electrochemical remediation is reasonably well developed; however, its design and implementation is so highly site-specific that specification of an off-the-shelf system with cleanup guarantees is not practicable.

Rather, the implementation of such a system must begin with a good understanding of the site and the contaminants. This information should be coupled with treatability studies using simple electrical cells and soil samples collected from the site. Conducting treatability studies on actual site samples is especially crucial for this technology since small differences in chemistry can have a major impact on process performance.

The vendor used for these treatability studies can be the supplier of the equipment or an independent party; however, a strong background in electrochemical processes and diffusion under electric fields is essential. The vendor should be capable of interpreting the results of the treatability studies and creating a set of performance specifications for the equipment. The equipment itself can be acquired through normal procurement channels. While competitive bidding is desirable, this application has not been applied extensively in the field, and ultimately, vendor selection should be based not only on a cost comparison, but also on an assessment of the vendor's experience with similar applications.

2.6.2 Start-up Procedures

Once installed, startup of an electrochemical system is relatively straightforward. After baseline contaminant concentrations have been established, the power is slowly ramped up to the pre-established design levels. A slow increase in power at startup insures that there are no unknown contingencies such as buried metal or pockets of salt that could short circuit the system. Injection of solutions into the site to improve the mobility of contaminants (i.e., chelating agents) should not be started until the system's performance without these agents has been established. Injection should also be gradually and electrosomotically made in increments to reach the target levels.

2.6.3 Operations Practices

No specific operation practices appear necessary for electrochemical processes other than maintaining liquid injection and withdrawal rates and power levels. However, groundwater levels and power usage should be monitored, and the process should be adjusted to reflect changes in these parameters over time.

2.6.4 Operations Monitoring

Monitoring programs for electrochemical systems should track two components: (1) the degree of concentration of the contaminants and pH and temperature changes at and around the electrodes, and (2) process conditions such as electricity usage, current, and potential.

For the second component, process conditions must be monitored to identify changes in the site over time. For example, under constant temperature, a change in current at constant potential (voltage) or potential at constant current could indicate a chemical change in the system. Power usage in a DC system is the product of current and potential.

2.6.5 Quality Assurance/Quality Control (QA/QC)

As with any remediation effort, QA/QC is an integral part of electrochemical process implementation. The QA/QC program should be developed on two general levels. The first is to meet the operating needs of the system. This should include the tests discussed in Section 2.5.10 which ascertain overall system performance.

The second level is intended to satisfy regulatory requirements; the QA/QC program must clearly demonstrate that the system is meeting all environmental and legal goals and requirements.

2.7 Case Histories

Both laboratory- and pilot-scale tests have proven electrochemical processes to be successful techniques for site remediation. The results of some of these test efforts are shown in Figures 2.7 through 2.9 and described in the following sections.

2.7.1 Laboratory-Scale Tests

At the laboratory-scale, electroosmosis has been applied to a wide range of soils and contaminants. Laboratory tests have shown effective cleanup is possible with electroosmosis, but it depends on many variables including types of contaminants, pH, initial concentrations, and adsorption and cation-exchange capacity of soil.

Figure 2.7
Bench-Scale Electrochemical Cell

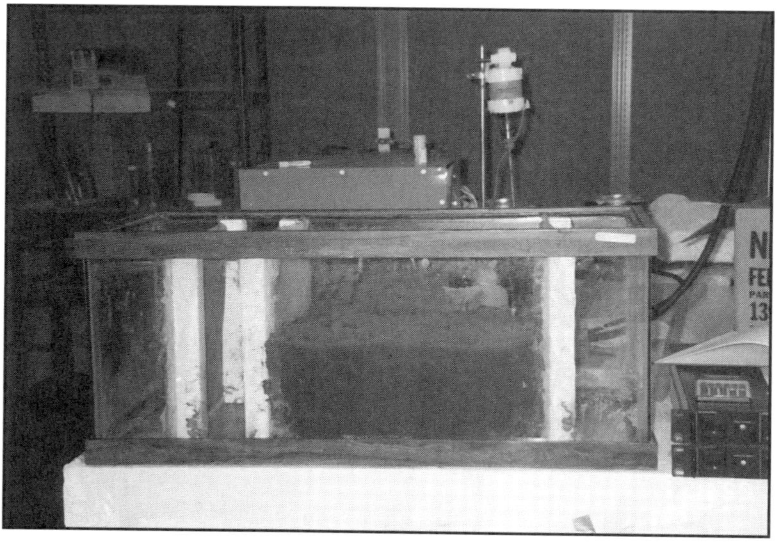

Reproduced courtesy of Electrokinetics, Inc. (1995)

Organic compounds are amenable to electroosmosis. Bruell, Segall, and Walsh (1992) achieved a removal of 15-25% for benzene, toluene, trichloroethylene, and m-xylene from kaolin clay in only 2-5 days of treatment. Shapiro and Probstein (1993) and Probstein and Hicks (1993) removed over 90% of phenol with an initial concentration of 450 mg/L from compacted kaolin clay samples by extracting less than 1.5 pore volumes — an indication that electroosmosis could be a very effective remediation method. A lower removal was observed at a lower initial concentration (45 mg/L) and was attributed to the adsorption of phenol on the clay. The effect of initial concentration on removal efficiency was more pronounced for acetic acid (Shapiro and Probstein 1993), where acetate ions electromigrated to the anode (the degree of dissociation of acetic acid is higher at a higher pH, which is caused by a low concentration of acetic acid).

Figure 2.8
One-Ton Soil, Pilot-Scale Electrochemical Cell

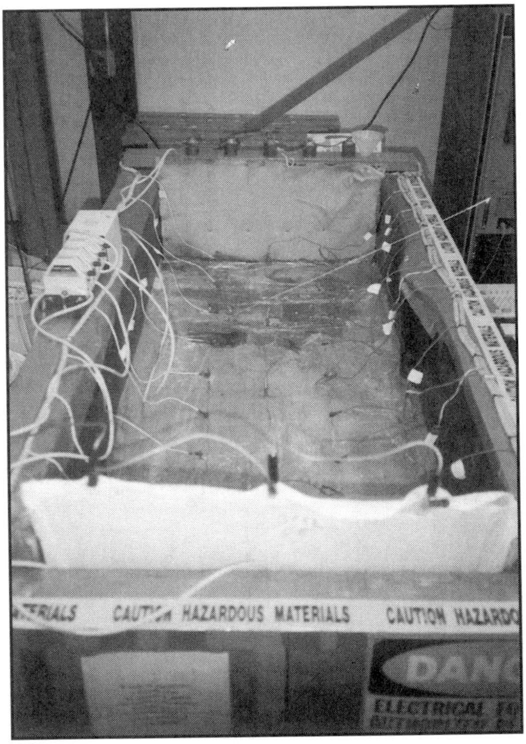

Reproduced courtesy of Electrokinetics, Inc. (1995)

For removal of metal ions, Pamukca and Wittle (1992) demonstrated, using laboratory-prepared synthetic samples, that 85 to 95% of cadmium, cobalt, nickel, and strontium were removed from commercially-obtained kaolinite and bentonite, prepared clayey-sand, and prepared/washed New Jersey beach sand. They attributed the removal to the movement of the acid front toward the cathode which caused dissolution of metal precipitates and the desorption of adsorbed metal ions. Of the soils tested, metals were the most difficult to remove from the bentonite. This is consistent with the higher ion-exchange capacity of montmorillonite, a major ingredient of bentonite, as indicated in Table 2.1. In addition, the treatment of clayey sand and bentonite was most influenced by the chemistry of the individual metal ions. The removal of metals is also affected by the buffering capacity (alkalinity) of soils, which resists a pH drop caused by an advancing acid front. The buffering capacity is due to the cation-exchange capacity and precipitates such as calcium carbonate (Acar and Alshawabkeh 1993). To dissolve metal precipitates in highly-buffered soils, additional hydrogen ions may be needed and can be introduced in the form of an acid solution at the anode.

Figure 2.9
Full-Scale Electroosmosis/Electromigration
Treatment System by Electrokinetics, Inc.

Light-weight HDPE liner material on surface reduces evaporation and escape of volatiles.

Reproduced courtesy of Electrokinetics, Inc. (1995)

Other ions were also found to be amenable to removal such as lead, manganese, and zinc (Khan and Alam 1993); copper and sulfate (Runnells and Wahli 1993); nitrate (Segall and Bruell 1992); uranium (Acar and Alshawabkeh 1993; Ugaz et al. 1994); and zinc (Probstein and Hicks 1993). The removal of radium and thorium was found to be poor due to the formation of insoluble precipitates in soil (US EPA 1995c; Ugaz et al. 1994).

2.7.2 Pilot-Scale Tests

Two companies conducted pilot-scale tests and cleanup projects: (1) Electrokinetics, Inc. (Baton Rouge, Louisiana) used 1-ton specimens of kaolinite and a mixture of fine sand and kaolinite for its pilot-scale tests (US EPA 1995c) and (2) Geokinetics (The Netherlands) reported on two field pilot-tests and three cleanup projects at various sites (paint factory, galvanizing plant, timber-impregnation plant, landfill, and military depot) covering a surface area of 50 to 2,800 m^2 (60 to 3,350 yd^2) and a depth of 1 to 2.6 m (3 to 8 ft)(Lageman 1993). The contaminants (all metals) and concentrations in the tests are listed in Table 2.2.

Table 2.2
Contaminant Concentrations for In Situ Electrochemical Pilot-Scale Tests

Contaminant	Concentration in mg/L (test)
Arsenic	400-500 (Geokinetics)
Cadmium	2-3,400 (Geokinetics)
Chromium	less than 300 (Geokinetics)
Copper	500-1,000 (Geokinetics)
Lead	100 to less than 5,000 (Geokinetics)
	850-5,322 (Electrokinetics)
Nickel	860 (Geokinetics)
Zinc	7,010 (Geokinetics)

Electrodes were separated by 0.7 m (2.3 ft)(Electrokinetics) and 1 m (3.3 ft) (Geokinetics). Electrokinetics applied a one-dimensional electric field for the tests, whereas Geokinetics used a system of horizontal cathodes buried 0.5 m (1.6 ft) below the ground surface and vertical anodes 1 m (3.3 ft) deep and 1 m (3.3 ft) apart). Reported ranges of power consumption were similar: 300 to 700 kWh/m^3 (230 to 540 kWh/yd^3)(Electrokinetics) and 65 to approximately 300 kWh/m^3 (50 to 230 kWh/yd^3)(Geokinetics). However, a value of as high as 800 kWh/m^3 (600 kWh/yd^3)was suggested to achieve a treatment goal in one Geokinetics test. Values of charge density reported were 133 $\mu A/cm^2$ (858 $\mu A/in.^2$)(Electrokinetics) and 400-800 $\mu A/cm^2$ (2,600 to 5,200 $\mu A/in.^2$)(Geokinetics).

Most tests resulted in successful remediation (70 to >90% removal) after one to several months of operation. The following suggestions and observations were noted:

- Electrokinetics suggested the depolarization of the cathode using acetic acid to prevent the precipitation of lead near the cathode;
- the buffering capacity (alkalinity) of soil influences the transport of metal ions. A low-pH soil (pH 4) facilitated the mobilization of lead at a low-energy dosage, whereas a highly-buffered soil retarded the mobilization of zinc requiring an additional energy input to lower the pH to between 3 and 4;
- metallic objects (>10 cm in size) in soil interfered with the removal by providing preferential paths for electrical current and should be removed in a pretreatment step along with insulating objects (plastic, wood, and concrete), whenever possible; and
- concretions of cadmium sulfide (a few mm to several cm in size) were found to prolong the removal. Two pretreatment methods were suggested: (1) the use of acid to dissolve cadmium from the concretions and (2) the removal of concretions by sieving.

In summary, the researchers found that effective cleanup of contaminated low-permeability soil is possible with electroosmosis/electromigration. However, many design and operational variables need to be considered for this technology to be cost-effective.

3

IN SITU PERMEABLE, ELECTROCHEMICALLY ACTIVE METAL BARRIERS

3.1 Scientific Principles

In situ permeable electrochemically-active metal barriers can treat groundwater contaminated with dissolved halogenated organic compounds and certain types of oxidized metals. The process is based on the fact that many common contaminants, both organic and inorganic, react with iron and other metals in their elemental (zero-valence) state. Conceptually, the treatment process is very simple, see Figure 3.1. A permeable barrier consisting of a trench or a wall structure filled with granular iron or other metal is placed in the flow path of the contaminated groundwater passing through the site. Contaminants in the water react with the metal and are reduced to less environmentally objectionable and more controllable forms as the water flows through the permeable barrier. This technology has been described in the literature by a number of different names such as porous-reactive walls, permeable walls, reactive iron walls, and permeable reaction walls.

The organic chemical reduction process is similar to the electrochemical processes previously discussed with the exception that the electrodes in those processes are replaced with a sacrificial metal such as aluminum, brass, copper, iron, or zinc. *Permeable barriers* can treat halogenated organic nitrates, organic nitrites and certain metals. Halogenated organic contaminants are dehalogenated. Metals, such as hexavalent chromium, are reduced to less hazardous or less soluble forms.

Figure 3.1
Schematic of Permeable Barrier

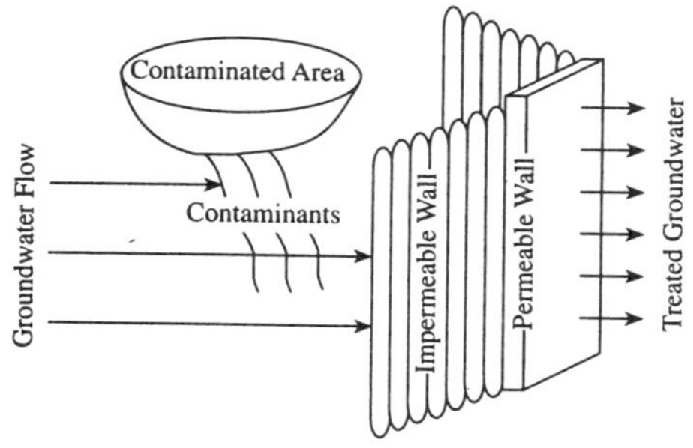

Because of its comparatively low price and ready commercial availability, iron has been the most widely-tested and used metal. Field trials have shown granular iron barrier walls to be successful in dechlorinating chlorinated organic compounds in groundwater. Laboratory data indicate that an iron barrier can also reduce hexavalent chromium to trivalent chromium. However, this reduction process may form precipitates which could cause significant plugging problems. When contaminated groundwater passes through this "permeable wall of iron metal" (O'Hannesin and Gillham 1992; Gillham and O'Hannesin 1994), the metal is oxidized, serving as an electrochemical galvanic half-cell, and contaminants and any other electron acceptors (e.g., O_2) are reduced to complete the electrochemical cell.

In field application, permeable barriers can be installed to either cover the entire plume of contaminated groundwater or intercept the plume using a "funnel-and-gate" approach with impermeable barriers depending on site conditions (Focht et al. 1996; Shoemaker et al. 1995). The permeable barrier consists of ground iron or another metal appropriate for the application. The finer the particle size of the metal, the greater the surface area to which the contaminated groundwater is exposed, and hence, the faster the rate of chemical destruction. However, with finer particulate, the permeability of the barrier decreases. Therefore, design requires testing to ensure a balance between these two characteristics. In either case, permeable barriers are subject to plugging as oxidation products, carbonates, and other precipitates accumulate in the pores.

When iron is selected as a sacrificial reactive metal, the following corrosion or electrochemical reactions take place within the barrier (Matheson and Tratnyek 1994; Wilson 1995; Sivavec and Horney 1995):

$$Fe \rightarrow Fe^{++} + 2e^{-} \tag{3.1}$$

At the same time, contaminants are reduced. An example of groundwater contaminated with a halogenated hydrocarbon and hexavalent chromium follows:

$$RX + H^{+} + 2e^{-} \rightarrow RH + X^{-} \tag{3.2}$$

$$Cr^{+6} + 3e^{-} \rightarrow Cr^{+3} \tag{3.3}$$

where: R = an alkyl group; and
X = a halogen.

The following are the overall chemical reactions:

$$Fe + RX + H^{+} \rightarrow Fe^{++} + RH + X^{-} \tag{3.4}$$

$$2Cr^{+6} + 3Fe \rightarrow 2Cr^{+3} + 3Fe^{++} \tag{3.5}$$

The trivalent chromium that is formed in the preceding chemical reaction is the relatively insoluble precipitate, chromium hydroxide. The aliphatic chlorides are converted to the relatively environmentally-benign and readily biodegradable aliphatic compounds. The chlorine in the aliphatic chlorides is converted to chloride.

The chemical reduction of many chlorinated organics using iron metal was found to follow first-order kinetics (O'Hannesin and Gillham 1992; Gillham and O'Hannesin 1994; Tratnyek 1966; Cipollone et al. 1997). Research and field results have shown that degradable hydrocarbons can be degraded to below detection limits given sufficient contact time.

One of the major factors that influence the electrochemical reduction is pH. As the reaction proceeds, the pH of water increases as shown in Equations 3.2 and 3.4. When the pH is greater than 8, the reaction slows substantially (Senzaki and Kumagai 1988, 1989; Senzaki 1991; Matheson and

Tratnyek 1994). The high pH also causes metals to precipitate, leading to the loss of porosity of the flow path. In the case of the reduction of hexavalent chromium, the production of chromium hydroxide is beneficial for immobilizing the chromium within the barrier; however, this same precipitation clogs the pores of the permeable barrier, and a concomitant decrease in the rate of water flow through the barrier.

For additional information on the use of semi-permeable metal barriers in the field and a general description of the chemical processes and some of the design considerations, the reader is referred to Matheson and Tratnyek (1994). For a more detailed discussion of the chemistry involved and of recent research and filed results, see Johnson, Shcerer, and Tratnyek (1996).

3.2 Potential Applications

If permeable barrier treatment is used as an alternate containment strategy, high permeability in contaminated soil is not required. However, if permeable barrier treatment is used as a primary strategy to clean up contaminated soil and groundwater, it requires relatively high permeability (a hydraulic conductivity of much higher than $1 \cdot 10^{-5}$ cm/sec) in the contaminated area so that contaminants can move out of the soil and through the permeable barrier. In addition, the treatment has to be designed such that contaminants do not escape the area without being treated. For example, the barrier needs to be more permeable than the surrounding formation to avoid the buildup of hydraulic pressure behind the barrier. The addition of impermeable barriers at strategic locations may be necessary depending on site conditions.

According to Vogan et al. (1995) and Gillham and O'Hannesin (1994), various halogenated hydrocarbons can be reduced by this form of treatment. Table 3.1 lists halogenated hydrocarbons that were tested for electrochemical reduction. Many of these hydrocarbons rapidly degrade with half lives ranging from a few minutes to a day where the iron surface area is 1 m^2/mL. Gillham and O'Hannesin (1994) compared the half lives from electrochemical reduction with those from natural abiotic degradation. As shown in Table 3.1, the natural abiotic degradation rates were many orders of magnitude slower than the electrochemical reduction rates.

Table 3.1
Halogenated Hydrocarbons Evaluated for Electrochemical Reduction Using Iron as a Sacrificial Metal

Compounds	Initial Concentration, (μg/L)	Half Life[1,5] Electrochemical-Reduction (hr)	Half Life[3,4] Natural Abiotic Degradation (hr)
	Methanes		
Dibromomethane, CH_2Br_2	–	degraded[2]	–
Dichloromethane (Methylene Chloride), CH_2Cl_2	–	no degradation[2]	–
Trichloromethane (Chloroform), $CHCl_3$	2013	1.49	$1.6 \cdot 10^7$
Tribromomethane (Bromoform), $CHBr_3$	2120	0.041	$3.6 \cdot 10^5$
Tetrachloromethane (Carbon Tetrachloride), CCl_4	1631	0.020	$6.0 \cdot 10^6$
	Ethanes		
Chloroethane, CH_3CH_2Cl	–	no degradation[2]	–
1,2-Dichloroethane, $ClCH_2CH_2Cl$	–	no degradation[2]	–
1,1,1-Trichloroethane, CH_3CCl_3	683	0.065	$9.6 \cdot 10^3$
1,1,2,2-Tetrachloroethane, $Cl_2CHCHCl_2$	2513	0.053	$3.5 \cdot 10^3$
1,1,1,2-Tetrachloroethane, Cl_3CCHCl	2334	0.049	$4.1 \cdot 10^5$
Hexachloroethane (Perchloroethane), CCl_3CCl_3	3621	0.013	$1.6 \cdot 10^3$
	Ethenes		
Chloroethene (Vinyl Chloride), CH_2CHCl	3663	12.55	–
1,1-Dichloroethene (Vinylidene Chloride), CH_2CCl_2	2333	5.47	$1.1 \cdot 10^{12}$
Trans-1,2-Dichloroethene, ClCHCHCl	1774	6.41	$3.9 \cdot 10^{14}$
Cis-1,2-Dichloroethene, ClCHCHCl	1949	19.7	$3.9 \cdot 10^{14}$
Trichloroethene, CCl_2CHCl	1555	0.67	$1.1 \cdot 10^{10}$
Tetrachloroethene (Perchloroethene), Cl_2CCCl_2	2246	0.28	$8.7 \cdot 10^{10}$
	Others		
Trifluorotrichloroethane, $C_2F_3Cl_3$ (Fluorocarbon-113)(FC-113)	–	degraded[2]	–

[1]Gillham and O'Hannesin (1994)
[2]Vogan et al. (1995)
[3]Jeffers et al. (1989)
[4]Vogel, Criddle, and McCarty (1987)
[5]Assuming that the barrier has an iron surface area of 1 m^2/mL.

Source: Gillham and O'Hannesin 1994; Vogan et al. 1995

3.3 Treatment Trains

An in situ permeable, electrochemically-active metal barrier treatment system needs to be designed such that contaminants and contaminated groundwater do not escape the area without passage through the permeable barriers. The installation of impermeable barriers may be necessary to funnel the groundwater to the face of the permeable barriers. The impermeable barriers can be sheet piling, grout curtains, or other standard designs appropriate for the site. Once the treatment system is installed, operation of the system is self-regulating; however, monitoring is required.

3.4 Remediation Goals

The general application for a permeable wall treatment system is similar to that for a pump-and-treat system including aboveground water treatment — to reduce the concentration of the target contaminant in the groundwater flowing through the walls to acceptable levels. Table 3.1 provides the initial concentrations and half lives of a variety of contaminants found in groundwater when passed through an iron bed.

3.5 Design

3.5.1 Design Basis

The design of a permeable treatment wall is mainly based on the groundwater velocity and seasonal direction, aquifer hydraulic conductivity, distribution of conductivity, and the concentration and distribution of contaminants of concern. The distribution of contaminant and the hydraulic conductivity determines whether a continuous permeable wall or funnel-and-gate arrangements are most cost-effective. The wall, optimally, is placed perpendicular to the plume center line, transverse to groundwater flow. Seasonal variation of the flow direction with respect to the wall is taken in account into designing the thickness of the wall. The wall either should extend vertically to an aquitard or flow modeling performed to design the depth to which the wall must be extended in the aquifer to capture the entire plume. Currently, walls are limited to a trenching depth of about 50 ft, perhaps 75 ft if sealable sheet piles can be driven to that depth (requires absence of boulders and limited friction with the soil involved). Groundwater models and wall porosity selections are used to design the width of a wall beyond plume limits to ensure complete capture. Wall permeability is a critical design parameter necessary to direct the complete contaminant plume into the reactive media. Generally, walls are constructed to be more permeable than the aquifer into which the wall is placed. Precipitation and clogging must be considered in designing the wall to be adequately permeable for the treatment period intended. Otherwise, the estimate of the present cost of the system should reflect the need to replace or regenerate the wall once clogging interferes with wall performance. Currently, a 5- to 10-year design life is assumed. It is best if a wall can be extended to an aquitard to prevent leakage of a contaminant under and around a wall. However, cost-effective, hanging walls can be designed if careful attention is given to porosity in designing and installing the wall. Wall thickness is based on the flow velocity through the media (including increased velocity if impermeable walls are used to direct flow), kinetics of degradation, the type of iron (i.e., granulated reagent grade, blast furnace dust, steel furnace dust) selected, and the surface area of the iron, as well as porosity. Highly oxidized waters may require a pretreatment through a vertical layer consisting of a mixture of granular cast iron and large size sand or pea gravel. The pretreatment layer distributes the flows and precipitates dissolved solids in a layer with a high void volume. High dissolved solids are generally not known to be a problem, but groundwater chemistry in the presence of zero-valent iron must be taken into account when designing wall porosity and thickness. High concentrations of nitrate, sulfate, carbonate, and other oxidized species are important because these species may interfere with the reduction of chlorinated compounds (i.e., TCE, PCE, vinyl chloride). Multiple species of chlorinated solvents can be degraded simultaneously. However, the wall thickness must be chosen considering the reaction kinetics, influent concentration, and desired effluent concentration of each species. In addition, it should be noted that as the parent species (e.g., TCE) degrades within the wall, it is likely that dehalogenation daughter products (e.g., DCE) will appear in some proportion to the parent concentration. The daughter product may react more slowly than the parent and become the limiting factor in overall treatment, particularly if the daughter product already is present in the influent. Since these reaction kinetics are to a degree interactive and quite complicated, many designers select wall thickness on the basis of laboratory treatability testing and/or field pilot results. Alternative mixtures of granular iron and pyrite or other compounds may be used to control pH and thus control precipitation of carbonates and reaction byproducts and the life of the reactive wall.

3.5.2 Design and Equipment Selection

The most important consideration in the design and equipment selection for permeable barrier installations is the materials of construction for the barrier walls and grout curtains. Permeable treatment walls are usually made of granular iron. The barrier walls in existing installations have been made from interlocking sheet piling. These walls, formed from commercially-available, corrugated metal are installed using standard construction techniques. Other types of barrier walls, such as slurry wall, jet grouted barrier, or any other types of impermeable barrier, can be used as appropriate.

An important consideration for permeable barrier treatment is the nature of the bottom barrier to groundwater flow. The existing field installations have taken advantage of impermeable clay lenses and other layers underlying the contaminated areas. If such impermeable layers do not exist, or if they do not form a continuous impermeable barrier to groundwater flow under the site, then use of grouting techniques, possibly combined with horizontal drilling methods, may be necessary to seal the bottom of the treatment region. The effectiveness of horizontal drilling methods for this application has not yet been demonstrated.

Another potential concern is the nature of the iron used for the reactive barrier wall. To date, all work has used high-grade iron containing known levels of impurities. Other sources of granular iron are available, such as blast furnace dust (a high-grade iron). The Air Force Armstrong Laboratory is in the process of developing a protocol for selection of reactive media that is useful in defining specifications for suitable granular iron requirements (McCutcheon 1996). Iron from these other sources can be significantly cheaper than the high-quality material currently used to date. However, such iron can be contaminated with high levels of toxic impurities such as lead, mercury, zinc, cadmium, selenium, and arsenic. These metals cannot be introduced into aquifers without extensive investigation of their mobility and toxicity. As operating experience with permeable treatment walls increases, there may be an incentive to reduce costs by using these less-expensive sources of iron, but care must be taken to ensure that iron from such sources does not inadvertently contaminate the site.

3.5.3 Process Modification

The physical layout and construction of permeable electrochemically-active metal barriers is largely determined by site conditions. The in situ treatment system must be constructed to funnel all contaminated water flow into the permeable treatment barrier. One area of flexibility in the design is the composition of the barrier. Most field applications of this technology to date have used granular cast iron. It is possible, however, to mix the iron metal with either an inert solid or with magnetite (Fe_2O_3). No specific data on the potential effect of this alteration were found. It is possible to mix the iron metal with sand, but this is mainly used for porosity control. Typically, the greater the dilution of the iron surface area, the lower the mass of iron and the larger and deeper (in the direction of groundwater flow) the bed needs to be. Given an influent concentration and iron with a certain reactivity, the important parameter in design of a granular iron treatment barrier is the mass of iron placed in the path of the influent groundwater. Sufficient iron must be placed so as to achieve treatment to the desired effluent concentration. How this iron is configured is somewhat irrelevant in terms of the reactions. It could be placed as pure iron in a single trench, mixed with sand and placed in a larger trench, or placed into the aquifer through other means such as jetting or deep soil mixing. Whichever method is most cost effective will depend on how much iron is needed (in terms of equivalent wall thickness) and how deep it must be placed. However, the statement that mixing the iron with sand may aid in porosity control (i.e., versus a pure iron-filled trench) is correct as is the statement that ferrous iron may assist in the degradation process if used as the mixing agent with zero-valent iron.

Use of an equilibrium model, MINTEQU-A (Felmy, Girvin, and Jenne 1984), indicates that dissolution of pure iron metal in some groundwaters could cause a sharp increase in pH which would then result in the precipitation of naturally-occurring dissolved solids. The precipitates might adversely affect the performance of a barrier treatment system by:

- the formation and accumulation of precipitates within the reactive zone which could reduce the porosity of the system and change its hydraulic properties;

- the precipitate could form impermeable blocks in the groundwater flow pathway, thereby diverting the groundwater away from the permeable wall; and
- the precipitates could deposit on the active surface and de-activate it.

Ongoing laboratory studies (Holser, McCutcheon, and Wolfe undated-a) indicate a mixture of 25% (by weight) granular pyrite (iron disulfide) and granular iron may control pH. The results of the tests using three different iron-to-pyrite ratios imply that a 9:1 mixture of iron to pyrite may improve the destruction of chlorinated organic compounds primarily through control of pH (see Figure 3.2).

Figure 3.2
Conversion of TCE in Columns Packed with Mixtures of Iron and Pyrite

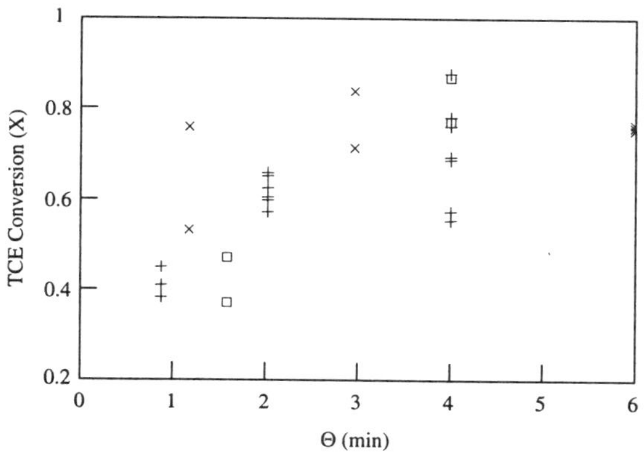

+ 50% Fe/50% Pyrite
× 75% Fe/25% Pyrite
□ 100% Fe

Mixtures of iron and pyrite were prepared and the pH of the effluent was monitored over a period of 500 residence times. The total amount of packing for each of these columns was 4.0 grams. TCE conversions were similar for the different columns despite the variation in the amounts of iron.

Source: Holser et al. undated-a

Other process modifications that may be required relate to the possibility of plugging or deactivation of the treatment medium. It is noted that no significant precipitates were observed in the in situ reactive wall at the University of Waterloo Borden test site two years after installation (Tratnyek 1996; Matheson and Tratnyek 1994). The wall has performed consistently for about 3.5 years with no noted problems. Data from in situ systems installed in California in December 1994 and September 1995 and from other in situ applications will generate further data to provide additional understanding of plugging and deactivation.

3.5.4 Pretreatment Processes

No pretreatment is required for application of this process.

3.5.5 Posttreatment Processes

If porosity is carefully designed to account for precipitation and clogging or pH is controlled to avoid precipitation, no posttreatment is expected to be required. However, there may be a need for periodic treatment of the reactive material in the permeable barrier wall to remove precipitates. No unacceptable quantities of precipitates were found at the Waterloo-Borden test site during the (approximately) first five years of operation. Subsequently, some severely-clogged column treatability tests indicate that clogging is a possible problem with some groundwaters. According to the manufacturer of that system, laboratory measurements of alkalinity losses indicate the possibility of a 2 to 15% loss in porosity per year. However, the vendor has suggested that the amount of precipitation that will occur in situ will be significantly less than predicted from laboratory studies due to the groundwater used for laboratory studies. He indicates that sampling and transport saturates the samples with oxygen and shifts the carbonate equilibrium in the groundwater used for laboratory studies. These changes, which are caused by the sampling and transport result in plugging of the iron column used for the laboratory tests.

3.5.6 Process Instrumentation and Control

The process requires no unique instrumentation or controls. Other than monitoring wells, the only addition to a site would be provisions for sampling at different levels up and downstream of the treatment system and continuous groundwater level monitoring. Water level monitoring is necessary to indicate a reduction in the barrier's permeability or occurrence of blocking that may be due to precipitation.

Monitoring is expected to be more expensive as compared to other remediation methods for two reasons. First, sampling at many locations and at multilevels is needed to monitor the performance of the treatment system because of the in situ nature of the treatment. Second, environmental regulators may require additional information to ensure the treatment works as expected because it is an emerging technology. With experience gained at various ongoing demonstrations, monitoring requirements should soon decrease, however.

3.5.7 Safety Requirements

The iron metal used in permeable barriers, especially if it is a finely-divided form, is a potential fire hazard. The metal will oxidize exothermally upon exposure to air. If the metal is placed in contact with combustible material, the mixture might ignite. However, once in-place in a wet environment, the risk of fire is minimal. The material safety data sheet (MSDS) for iron supplied by the vendor should be consulted for proper handling procedures. Other sources of iron mentioned previously would avoid this hazard. In addition, the formation and accumulation of hydrogen gas from hydrogen ions is possible. Therefore, all monitoring wells will have to be designed to vent all gases to prevent hydrogen gas buildup.

3.5.8 Specification Development

The specifications for installation of a permeable barrier treatment system focus on on-site geology. It is imperative that the location of permeable and impermeable zones in the area be well known so that appropriate impermeable barriers can be installed to funnel the contaminated groundwater flow into the treatment barrier. Since the cost of construction and placement of the barrier is the greatest expense in the implementation of this technology, relatively small variations in the site's physical characteristics can have a major impact on the system cost.

3.5.9 Cost Data

The overall cost of a permeable electrochemically-active metal barrier system include:

- site survey and detailed geochemical assessment costs;
- installation costs for impermeable barriers and for the permeable barrier treatment system; and
- monitoring costs.

Since the system is passive, once installed, there are no unique system operating costs. Monitoring is common to any in situ remediation system, but as indicated earlier, because of the developing nature of this technology, additional monitoring may be required.

Cost data provided by the vendor of this technology (Vogan 1996) follow. The key parameter in determining costs is the dimension of the in situ treatment system. The material and construction costs for a 180 m (600 ft) wide, 9 m (30 ft) deep funnel-and-gate system to treat 50 to 60 mg/L trichloroethylene (TCE), 10 mg/L perchloroethene (PCE), and 20 mg/L trans-1,2-dichloroethene (DCE) in a high-velocity aquifer were estimated at about $1.5 million. Costs to treat a narrower plume (30 m [100 ft]) of similar depths containing 10 mg/L PCE and 2 mg/L TCE were estimated at $950,000. Costs to treat a relatively narrow (30 m [100 ft]) and shallow (8 m [25 ft]) plume containing several mg/L TCE were estimated at $275,000. These are material and installation costs and do not include costs for engineering, licensing fees, soil disposal, or health and safety measures, which could easily add 30 to 50% to these estimates.

The main cost unique to this technology is the cost of iron. Granular cast iron has an approximate bulk density of 0.112 tonne (0.1 ton)/ft^3. One vendor of granular iron suitable for this application (Master Builders) quoted a price of $500 to $666/tonne ($450 to $600/ton) depending on quantity and purity. Focht et al. (1996) quoted even a lower price of iron [$560/tonne ($400/ton)]. These prices are substantially lower than these assumed by Vogan (1996) $833/tonne ($750/ton) in the above cost analysis. Therefore, the construction cost estimates provided by Vogan may be somewhat high.

The only routine operating costs for a properly-functioning system are those associated with monitoring. However, should plugging occur it may become necessary to flush the permeable wall system with fluids to remove precipitates from the reactive media. No information on the costs of such regeneration were found. Should flushing prove unsuccessful in clearing the blockage, reactive media removal and replacement will be necessary. The cost of such replacement depends on the quantity and cost of the iron.

Because of the high front-end and low operating costs of the permeable reaction wall treatment process compared with those for competing processes, the cost analysis for any application must be performed on a present-worth (or annual cost) basis. Costs supplied by EnviroMetal Technologies, Inc. indicate that the costs can be lower for this process than for competing processes when treating groundwaters contaminated with volatile organic compounds (although no underlying assumptions were given). Also, the exact cost of the permeable barrier treatment portion of the overall treatment systems quoted above appears to be relatively small. The costs quoted are:

Industrial facility, California (3 yr net present value)

- EnviroMetal process — $2.9M
- Pump-and-treat — $7.8M
- Dewater, soil vapor extraction — $4.1M

Somersworth Sanitary Landfill, New Hampshire

- EnviroMetal Process

Installation	$12.74M
Operation	$2.22M
Total	$15.0M

Of this total, EnviroMetal process capital costs amount to about $1.5M

- Pump-and-treat alternative with impermeable cap, etc.

Installation	$16.5 to 18.4M
Operation	$2.8 to 3.2M
Total	$19.7 to 21.2M

Landfill, New York

The present-worth values reported below supposedly include other aspects of the site remediation, such as monitoring, soil treatment, etc., however, details were not provided.

	Total Present Worth	O&M Present Worth
• EnviroMetal process	\$2.7M	\$0.6M*
• Air sparging	\$2.5M	\$0.7M*
• Air sparging/in situ bio	\$2.9M	\$0.9M
• Extraction (pump)	\$2.9M	\$0.8M

*Of these installation costs, the EnviroMetal process involved about \$500,000 in capital costs.

Industrial Facility, midwest U.S.

• EnviroMetal process	\$0.7M
• Soil treatment and pump-and-treat	\$0.8M
• Soil excavation and pump-and-treat	\$1.5M

Note: The EnviroMetal process is a proprietary permeable reactive barrier process marketed by EnviroMetal Technologies Inc.

3.5.10 Design Validation

The principal design validation required is regular monitoring of the groundwater before and after the treatment barrier and continuous monitoring of the groundwater level at points upstream of the groundwater barrier.

3.5.11 Permitting Requirements

There appears to be no unique permitting requirements for a reactive barrier treatment system beyond approval of the remediation plan required for all types of site cleanups. The technology is passive, it has no NPDES discharges, produces no hazardous waste that must be disposed or further treated, and it has no point sources of air emissions that might require permitting.

3.5.12 Performance Measures

The process has no unique measure of performance beyond the results of continued monitoring through sampling and analysis.

3.5.13 Design Checklist

The process is too new to enable providing a meaningful design checklist.

3.6 Implementation and Operation

3.6.1 Implementation Strategies

The implementation of such a system must begin with a good understanding of the site and the contaminants. This information should be coupled with treatability studies using groundwater samples collected from the site. Conducting treatability studies on actual site samples is especially crucial for this technology since small variations in chemistry can have a major impact on process performance.

The vendor used for these treatability studies can be the supplier of the equipment or an independent party; however, a strong background in electrochemical processes is essential. The vendor should be capable of interpreting the results of the treatability studies and creating a set of performance specifications for the equipment. The equipment itself can be acquired through normal procurement channels. While competitive bidding is desirable, this application has not been applied extensively in the field, and ultimately, vendor selection should be based not only on a cost comparison, but also on an assessment of the vendor's experience with similar applications.

Although the basic research on permeable barriers appears to have been conducted at U.S. and Canadian facilities, certain components of the technology may be proprietary to EnviroMetal Technologies, Inc. Legal guidance should be sought to avoid possible patent or other infringements.

3.6.2 Start-up Procedures

The barrier treatment process is a passive system. Once installed, it requires no "start-up", and operates without the need for adjustment. Compliance monitoring should be delayed until one or two pore volumes (volume in the soil available for liquid or gas) has passed through the barrier.

3.6.3 Operations Practices

After installation, the operation of a permeable barrier treatment system involves monitoring and correction of blockages if they occur.

3.6.4 Operations Monitoring

Regular monitoring of the groundwater before and after the treatment barrier and continuous monitoring of the groundwater level at points upstream of the groundwater barrier are required.

3.6.5 Quality Assurance/Quality Control (QA/QC)

Standard QA/QC procedures related to permeable barrier treatment need to be followed for all analyses. Successful system installation requires that the permeable wall be homogeneous so that the groundwater flow takes advantage of all of the system's reactive volume. In addition, care must be taken to ensure that impermeable barriers to flow are not inadvertently breached.

3.7 Case Histories

Permeable barrier treatment has been the subject of several studies at various scales. This section describes two programs — one system at the pilot-scale and a second, which is ongoing, at full-scale.

Gillham (1995) described a study at the Canadian Forces Base Borden field site where a 1.4 m (4.5 ft) thick permeable wall was used that contained 22% (by weight) of granular iron and 78% (by weight) of sand. The residence time of groundwater in the wall was 16 days at a flow rate of 9 cm/day (3.5 in./day). The contaminants were trichloroethene and perchloroethene at concentration of 270 mg/L and 43 mg/L, respectively. As the groundwater passed through the wall, the concentrations decreased by 90 and 88%, respectively. During contaminant degradation, the production of trans-1,2-dichloroethene and 1,1-dichloroethene was observed. These byproducts represented less than 10% of the trichloroethene and perchloroethene removed and were found to be degraded. All the degradation of the chlorinated hydrocarbons was attributed to electrochemical reactions; no appreciable biological degradation of these compounds was found.

Yamane et al. (1995) and Wilson (1995) described the installation of a permeable barrier at a former semiconductor manufacturing facility (Intersil) in Sunnyvale, California, to replace an above-the-ground treatment facility. The site was contaminated with 1,2-dichloroethene, trichloroethene, CFC 113, and vinyl chloride. The barrier contained 100% iron and was 4 m (13 ft) deep and 1.2 m (4 ft) wide with two slurry walls at the ends of the barrier to guide the contaminated groundwater. The residence time of the groundwater in the barrier was greater than 2 days. The cleanup standards to be met were equivalent to the maximum contaminant levels in groundwater set by the California Department of Health Services or the US EPA. The residence time of 2 days was selected based on the degradation of vinyl chloride, which degrades slowly and has a long half-life as shown in Table 3.1.

4

SUPERCRITICAL WATER OXIDATION

4.1 Scientific Principles

Supercritical water oxidation (SCWO), also referred to as hydrothermal oxidation (HTO), is a technology for the destruction of hazardous and non-hazardous wastes. Relative to conventional incineration technologies, it offers the following advantages:

- equivalent levels of organic destruction;
- potential for complete containment of all effluents until acceptable treatment has been verified;
- lower temperature oxidation;
- negligible likelihood of the formation of dioxins or furans;
- ability to physically remove normally-soluble salts from an aqueous solution;
- enhanced process stability;
- negligible NO_x and SO_x production;
- negligible airborne particulates;
- compact equipment; and
- minimal pollution abatement equipment required.

SCWO destroys organic materials using an oxidant (usually air, oxygen, or hydrogen peroxide) in water above its critical point of 374°C (705°F) and pressures of 22.1 MPa (3,205 psi, 218 atm). Oxidation occurs under near-homogenous, single-phase conditions, which provide excellent mixing, high mass, and heat transfer rates. The organic destruction occurs in a relatively small volume reactor. Typical products from a SCWO process include carbon dioxide, water, nitrogen, metal oxides, and inorganic salts.

A common criticism of supercritical water oxidation is that the high pressures make the process dangerous. As explained in detail in Section 4.5.2, this is not true. Because of the rapid destruction times (on the order of 5 to 60 seconds) and high density of the material being treated, the actual volume of the reactor and ancillary piping and equipment under high pressure is very small. For example, a 76 L/min (20 gal/min) system typically requires a reactor volume on the order of 6 to 76 L (15 to 20 gal). This small volume can be safely maintained at the required pressures if materials of construction are appropriately selected.

Conceptually, SCWO is similar to wet oxidation [also called wet air oxidation (WAO)], a technique that has been used to treat sewage sludge for nearly one hundred years (Gloyna and Li 1993). WAO is discussed in another monograph of this series, *Thermal Destruction.* The difference between the two is that a SCWO system operates in the supercritical region of water, where water, organic materials, and oxygen are miscible forming a single phase. In water's subcritical region (where WAO systems operate) the water exists in both a liquid and gas phase.

The *critical point* of an element or compound is defined as the temperature and pressure at which the liquid and gaseous phases merge. Phase-change properties, such as heat of vaporization cease to have a meaning in the supercritical region. Therefore, for the purposes of this discussion, the material in the supercritical region is referred to as a "fluid" to differentiate it from liquids and gases.

In a SCWO system, the single-phase fluid greatly increases mass transfer rates between the oxygen and the organic compounds enabling high organic compound conversion levels with short residence times. A SCWO system can accomplish the desired treatment in times ranging from seconds to one or two minutes as compared to the many minutes or even hours required by a subcritical WAO system. A second important consideration when operating in the supercritical region is that normally-soluble inorganic salts are highly insoluble in the supercritical water. Removal of this salt is a design consideration. A variety of techniques offer potential solutions for removal of these salts from supercritical water.

As previously mentioned, the key parameter that differentiates SCWO from WAO is that a SCWO system operates above the critical temperature and pressure of water, defined above. Operation in the supercritical region tends to create a relatively homogeneous, nearly single-phase reaction medium between the oxygen and the organic materials to be oxidized. Therefore, the general chemical reaction for the destruction of the organic constituents can be reasonably well-defined as follows:

$$C_cH_hN_nP_pCl_lS_s + O_2 \rightarrow CO_2 + H_2O + HCl + H_3PO_4 + N_2 + H_2SO_4 \qquad (4.1)$$

Intermediate products such as organic acids may form, but they are also oxidized at SCWO reaction conditions. As indicated by Equation 4.1, the SCWO environment can be highly corrosive. The combination of oxygen and inorganic acids requires the various corrosion resistant materials of construction, such as nickel alloys, titanium, and platinum liners. Some designs minimize the need for such corrosion resistant materials by injecting a non-corrosive liquid stream that forms a protective layer over the system's surfaces.

Several generalized kinetic models, based on simplified reaction schemes involving the formation and destruction of rate-controlling intermediates, are available (Li et al. 1991). It is possible for the formed intermediate products to be destroyed within seconds, depending on the reaction rates. These rates are controlled by the operating temperature. If intermediate products are detected in the effluent, the temperature can be raised or the flow rate decreased to increase the residence time. Table 4.1 (Gloyna and Li 1993) lists the rate-controlling intermediates and the oxidation endproducts from several laboratory studies of various compounds at WAO and SCWO conditions. The kinetic parameters for the assumed first-order reaction are also given. As the rate parameters for the hydrocarbons and oxygenated compounds indicate, the reaction rates are substantially higher for the SCWO than for the WAO systems.

Reaction mechanisms and byproduct analyses indicate that short-chain carboxylic acids, ketones, aldehydes, and alcohols are the major oxidation intermediates under WAO conditions (Bailod, Faith, and Masi 1982). Kinetic studies of refractory compounds under SCWO conditions, such as acetic acid (Lee 1990), methanol (Rofer and Streit 1989), ammonia (Webley, Tester, and Holgate 1991), and carbon monoxide (Helling and Tester 1987) are well documented. For nitrogen-containing organic compounds, nitrogen

Table 4.1
Kinetic Parameters for Key Rate-Controlling Intermediates

Organic Compound Category	Key Intermediate (Alternative)	Oxidation End Product	Kinetic Parameters* Condition (Water)	k (1/sec)	E_a (kJ/mol)	Reference
Hydrocarbons and Oxygenated Hydrocarbons	CH_3COOH	CO_2, H_2O	Subcritical	$4.40 \cdot 10^{12}$	182	Foussard, Debellefontain, and Besombes-Vailhe 1989
			Supercritical	$2.55 \cdot 10^{11}$	172.7***	Wightman 1981
			Supercritical**	$2.63 \cdot 10^{10}$	167.1	Lee 1990
Nitrogen-Containing Organics	NH_3	N_2, H_2O	Supercritical	$3.16 \cdot 10^6$	157	Webley, Tester, and Holgate 1991
	(N_2O)			–	–	–
Chlorinated Organics	CH_3Cl	HCl, H_2O	Supercritical	–	–	–
	(CH_3OH)			$2.51 \cdot 10^{24}$	395	Rofer and Streit 1989

*Pseudo first-order reaction model using oxygen.
**Using hydrogen peroxide.
***Obtained from fitting the reported seven data points.

Reprinted from *Proceedings of the Second International Symposium on Environmental Applications of Advanced Oxidation Technologies* , Gloyna and Li, "Supercritical Water Oxidation: An Engineering Update," 1993 with permission of EPRI.

gas is generally the predominant SCWO end product regardless of the oxidation state of nitrogen in the contaminated material treated (Killilea, Swallow, and Hong 1992). Ammonia and nitrous oxide are formed under a variety of operating conditions (Shanableh 1990; Killilea, Swallow, and Hong 1992).

The heat to bring a waste stream to the temperature required for the SCWO process is derived from three sources: (1) heat released by the oxidation of the organic contaminants in the stream being treated, (2) heat released from materials added to the stream being treated, and (3) external heat supplied to the reactor or the feed streams. If the organic concentration in the feed stream is high enough (usually on the order of 10%) then the reaction is autogenous and external heat may be unnecessary. If not, then external heat will be necessary. It is necessary, therefore, to calculate the heating value of the organic constituents of the waste stream in order to specify the SCWO system.

The heat released from the oxidation of the organic compounds during the SCWO process is approximately equal to the Higher Heating Value of each of the constituents in the waste stream. The Higher Heating Value for many substances can be found in numerous combustion handbooks, such as Gill and Quiel (1993). As an alternative, heat release can be calculated based on the heat of formation of the constituents. This latter method allows for more precise calculation since it considers the heat consumed in the formation of the products of the oxidation reaction such as hydrochloric, sulfuric, and phosphoric acids.

The heat of reaction is calculated by (1) summing the heats of formation of the products of the chemical reaction, and (2) subtracting the sum of the heats of formation for the reacting species. The heat of formation is defined as the enthalpy change occurring during a chemical reaction where 1 mol of a product is formed from its elements. The heat of reaction, ΔH_r, is calculated from the heat of formation using the following formula:

$$\Delta H_{r,298K^\circ} = \Sigma n_p(\Delta H_f)_p - \Sigma n_r(\Delta H_f)_r \quad (4.2)$$

where: H_f = the heat of formation of the products and reactants (subscripts p and r, respectively) and the subscript 298°K (25°C, 77°F) refers to the reference temperature for the reactants.

Gill and Quiel (1993) include an extensive list of heats of formation for most organic and inorganic compounds commonly encountered in environmental situations. The heat of formation of most organic and inorganic compounds can also be found in chemistry and chemical engineering handbooks as well as in standard thermodynamic references. The effect of pressure on the heat of reaction in the liquid phase is relatively small and can be ignored in the preliminary design of a SCWO system.

The following example indicates how the heat of reaction of 1,3-dichloropropane (liquid) is calculated from the heat of formation. The heat of formation for 1,3-dichloropropane is 615 Btu/lb, and its molecular weight (MW) is 113. The oxidation equation is as follows:

$$C_3H_6Cl_2(L) + 4O_2 \rightarrow 3CO_2 + 2H_2O(L) + 2HCl \qquad (4.3)$$

The heat of formation for each of these compounds is as follows:

	$C_3H_6Cl_2$(L)	O_2(G)	CO_2(G)	H_2O(L)	HCl
Btu/lb	-615	0	-3,848	-6,832	-1,088
MW	113	32	44	18	36.5
Btu/lb-mol	-69,480	0	-169,294	-122,971	-39,713

Therefore, the heat of combustion of 1,3-dichloropropane liquid is:

$$\begin{aligned} &3(-169,294) + 2(-122,971) + 2(-39,713) - (-69,500) \\ &= -763,769 \text{ Btu / lb - mol} \\ &= \text{-6,759 Btu / lb} \end{aligned} \qquad (4.4)$$

In other words, 6,759 Btu are released when 1 lb of liquid 1,3-dichloropropane is oxidized, and the water resulting from the chemical reaction is condensed after forming.

4.2 Potential Applications

SCWO is applicable to the treatment of aqueous wastes and streams containing organic compounds including highly energetic and toxic materials such as propellants, explosives, or chemical warfare agents and their products of neutralization. It can also remove soluble inorganic compounds from wastewaters, and at the same time, destroy the organic constituents. While the SCWO system can be applied to dilute aqueous streams (i.e., those with a relatively low heating value), the cost of heating the stream to the required supercritical water temperature is significant.

SCWO is most appropriate for treating concentrated aqueous solutions of organic compounds where the heat of oxidation of the organic compounds can provide a significant fraction of the heat needed to operate the system. Assuming no energy recovery, the system becomes autogenous (i.e., all of the required heat can be provided by the chemical reaction, and the system becomes self-sustaining) at approximately 10-15% organics, depending on the heating value of the total organic content (Killilea 1996). Inclusion of a heat recovery system can lower this value significantly. The process has been applied to sludges such as those from wastewater treatment, but engineering problems can occur when pumping abrasive materials such as soil/water mixtures at the high pressures required by the system. SCWO is at present being considered as a serious alternative to incineration for the treatment of hydrolyzed propellant waste, hydrolyzed chemical agent (i.e., nerve gas, mustard gas), paper mill wastes, and other hazardous wastes. Section 4.7 describes a number of case histories of sites where SCWO has been identified as a potential solution to difficult contaminant destruction problems.

4.3 Treatment Trains

Figure 4.1 is a simplified schematic of the treatment train incorporating a basic SCWO system. Water and fuel can be added to the waste to be treated to decrease or increase its heating value, respectively. Water is added only if the organic content of the waste is higher than needed for an autogenous process. If the waste has too little organic material to maintain the desired operating temperature, fuel oil (kerosene or #2 heating oil) or other organic constituents (e.g., methanol) can be added to the waste stream to increase the overall amount of heat released and reduce the need for external heating. Alternatively, additional external heating is often preferred. If the waste contains halides (e.g., organic chloride) sulfur, or phosphorus, caustic additives can be added to neutralize the acids (HCl, HF, H_2SO_4, H_3PO_4) that form in the chemical reaction. Alternatively, acidic products might not need to be neutralized if the system's materials of construction can withstand the corrosive effects of the acids.

Figure 4.1
Simplified Process-Flow Diagram of SCWO

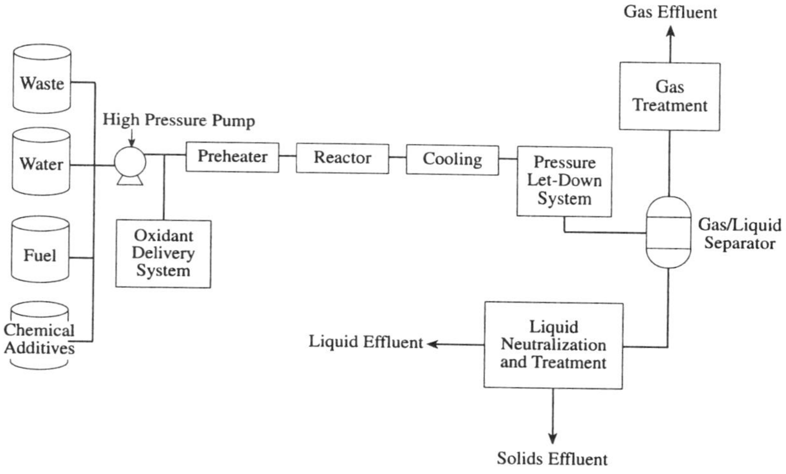

Reproduced courtesy of General Atomics

Once premixed, the waste stream is pressurized, preheated, and mixed with an oxidant such as compressed air, oxygen, or hydrogen peroxide. The mixture is then fed into the reactor.

The fuel added to the reactor reduces the amount of heat needed from outside sources; however, it increases the amount of oxygen that must be injected into the system. The decision of whether to inject additional fuel and oxygen into the reactor or to introduce the heat externally depends on the nature of the waste and on the relative cost of the two options.

The treated supercritical effluent from the reactor must be cooled and reduced in pressure prior to discharge or further treatment. Several methods of cooling may be used. The simplest cooling method is to "quench" cool, whereby the effluent is passed through a pressure reduction valve and mixed with cold water. This method of cooling is rapid and it minimizes or even eliminates the deposition of precipitates on equipment. A second means of cooling is by the use of heat exchangers. The working fluid which absorbs the heat from the supercritical effluent might be an external fluid, such as non-contact cooling water, or it might be the incoming fluid to the SCWO reactor. In the former case, a heat exchanger could be used to make process steam. In the latter case, the waste heat from the effluent is used to preheat the influent with a commensurate reduction in the system's energy requirement. The design tradeoffs between these approaches need to be examined from the perspective of economics and of the potential of a specific waste to foul the heat transfer tubes. The need for treatability studies to make this determination is readily apparent.

Pressure let-down can similarly be performed by various means. The most simple is by passing the effluent through an expansion valve. This has the inherent advantage of low cost, but the behavior of solids under the high turbulence of the expansion valve must be evaluated. Erosion of the valve is virtually unavoidable, although proper design and choice of materials can keep the level of erosion within acceptable limits. It is possible to use turbines for pressure let-down and, thereby, to recover some of the mechanical energy used for pumping; however, discussions with various practitioners have led to the conclusion that this is presently not a common practice, largely because the cost of pumping is a relatively minor component of the overall cost of treatment (this is discussed in greater detail later in this chapter) and because the corrosive environment and the potential for plugging and erosion reduce the turbine's life to the point where the cost of energy saved is less than the cost of the equipment and its maintenance.

After pressure let-down, the effluent usually enters a knockout drum where the gases are separated from the liquid. The liquids may be subject to further treatment prior to discharge. The gases, which consist mostly of excess oxygen, nitrogen, carbon dioxide, and water vapor, are passed through condensers and demisters before discharge to the atmosphere. Further gas treatment is usually unnecessary.

Table 4.2
SCWO Destruction Efficiency for Selected Organic Compounds

Compound	Temp† (°C)	Time† (min)	Concentration in (mg/L)	Concentration out (mg/L)	Destruction Efficiency† (%)
2-Butanone	400	5	6,210	251	95.96
	400	10	6,210	197	96.83
	450	5	5,140	136	97.35
	500	5	6,210	71	98.86
Methyl Ethyl Ketone (MEK)	400	10	5,140	273	94.68
	450	5	5,140	136	97.35
p-Chlorophenol	450*	3	1,000	<0.1	>99.99
o-Cresol	400	1.2	10,040	4453	55.7
	400	10	10,040	71.9	99.3
	450	1.2	10,040	1302	87.0
	450	10	10,040	20.9	99.8
	500	1.2	10,040	511	94.9
2,4-Dichlorophenol	400	10	300	1.2	99.6
	450	5	300	0.9	99.7
	500	5	500	1.6	99.7
	450*	4	1,000	<0.01	>99.999
Diethylene Glycol Diethyl Ether	450*	2	1,000	<1.0	>99.9
2,4-Dinitrotoluene	410	3	84.0	14.0	83.0
	528	3	180.0	<1.0	>99.0
	450*	1	200.0	<1.0	>99.5
Ethylene Glycol	450*	2	1,000	<1.0	>99.9
Pentachlorophenol	400	2	500	<0.04	>99.99
	450	2	500	<0.04	>99.99
	500	2	500	<0.04	>99.99

Table 4.2 cont.
SCWO Destruction Efficiency for Selected Organic Compounds

Compound	Temp† (°C)	Time† (min)	Concentration in (mg/L)	Concentration out (mg/L)	Destruction Efficiency† (%)
Pyridine	400	5	500	352.6	29.5
	450	10	500	4.1	99.2
	500	5	1,000	24.3	97.6
	500	20	500	1.8	99.6
Trichloroethylene	450	1	1,827	32	98.2
	450	5	1,827	13	99.3
2,4,6-Trichlorophenol	500*	2	200	<0.01	>99.995

All SCWO data were obtained from the batch tests with excess oxidant loading.
Pressure for all SCWO tests was about 27.6 MPa (4000 psi).

*Hydrogen peroxide was used; and oxygen was used for all other tests.
†These results are for systems operating at the low end of SCWO temperatures. Reactors operating in the 600-650°C range and less than one minute residence time routinely achieve 99.999% destruction efficiencies for all of these, and other compounds. Please see Section 4.7 for results of treatment at these elevated temperatures.

Reproduced with permission of the American Institute of Chemical Engineers from *Environmental Progress,* Volume 14, Number 3, Gloyna and Li, "Supercritical Water Oxidation Research and Development Updates," p 185. Copyright ©1995 AIChE. All rights reserved.

4.4 Remediation Goals

The remediation goal of supercritical water oxidation is to destroy the organic materials in an aqueous stream or sludge and, if necessary, concentrate the inorganic materials in a small-volume side stream. Very high levels of destruction of organic contaminants can be achieved; total organic carbon reductions of >99.999% have been reported for SCWO systems. Table 4.2 presents destruction data on selected compounds at various residence times and temperatures. Table 4.3 reports SCWO system performance on a variety of industrial organic wastes.

Table 4.3
SCWO Destruction Efficiency for Selected Organic Wastes[1,2]

Compound	Temp[5] (°C)	Time[5] (min)	Concentration in (mg/L)	Concentration out (mg/L)	Destruction Efficiency[5] (%)
Industrial Wastewater[3]	400	1	1,840	27	98.5
	450	1	1,840	15	99.2
	500	1	1,840	4	99.7
Industrial Sludge[4]	400	30	30,300	120	99.6
	450	10	30,300	50	99.8
	450	5	30,300	400	98.7
Mixture of Industrial	400	4	39,000	4,520	88.4
Wastewater and Sludge[4]	450	4	39,000	831	97.9
	500	4	39,000	429	98.9
Municipal Sludge[4]	400	8	14,020	687	95.1
	450	4	14,202	84	99.4
Contaminated Soils[4]	400	8	170	13	92.4
	450	4	170	9	94.6
	500	2	170	9	94.6

[1]All SCWO data were obtained from the batch tests with at least 100% excess oxygen.
[2]Test pressure was about 27.6 MPa (4000 psi).
[3]Total organic carbon.
[4]Chemical oxygen demand.
[5]These results are for systems operating at the low end of SCWO temperatures. Reactors operating in the 600-650°C range and less than one minute residence time routinely achieve 99.999% destruction efficiencies for all of these, and other compounds. Please see Section 4.7 for results of treatment at these elevated temperatures.

Reproduced with permission of the American Institute of Chemical Engineers from *Environmental Progress,* Volume 14, Number 3, Gloyna and Li, "Supercritical Water Oxidation Research and Development Updates," p 188. Copyright ©1995 AIChE. All rights reserved.

4.5 Design

4.5.1 Design Basis

The key to successful SCWO process design is the integration of various unit operations. Important design considerations include (Gloyna and Li 1993):

- reactor residence times and associated temperatures;
- reactor and ancillary equipment configuration;
- system pressures and related temperatures;
- materials of construction for each unit operation;
- control and removal of solids either from the supercritical fluid or the treated effluent; and
- operation and maintenance of the facility, including safety, analytical support, regulatory monitoring, and disposal requirements.

Generally, SCWO research has covered such areas as chemical reaction mechanisms and kinetics, salt formation and solubility, mass and heat transfer, transformation product identification, corrosion, catalysts, and additives. Process development has focused on materials of construction, reactor design, heat exchange and recuperative heat recovery, solid-liquid separation, gas-liquid separation, control systems, effluent handling, ash disposal, safety requirements, and process system integration.

A SCWO reactor of tubular design behaves like an ideal plug flow reactor as defined by Levenspiel (1962). Therefore, the kinetics of the reaction are a necessary consideration in its design. Table 4.4 lists the various global kinetic models for common SCWO reactions including kinetic parameters and waste types. The experimental conditions for each reaction are also included.

Table 4.4
Global Kinetic Models for Supercritical Water Oxidation of Organic Substances

Compounds	Oxidant	Reactor Type	Kinetic Parameters* $k^{\bullet}$	E_a	m	n	Temperature (°K)	Pressure (atm)	$[C_A]_o$ (g/L)	References
Acetamide	H_2O_2	flow	$2.75 \bullet 10^5$	88.3	1.15	0.05	673-803	240-350	1.5-4.0	Lee 1990
Acetamide**	H_2O_2	flow	$5.01 \bullet 10^4$	94.7	1	0.17	673-803	240-350	1.5-4.0	Lee 1990
Acetic Acid	H_2O_2	flow	$2.63 \bullet 10^{10}$	167.1	1	0	673-803	240-350	1.3-3.3	Lee 1990
Acetic Acid	H_2O_2	flow	$9.23 \bullet 10^7$	131	1	0	673-773	240-350	1.0-5.0	Wilmanns 1990
Acetic Acid	O_2	flow	$9.82 \bullet 10^{17}$	231	1	1	611-718	394-438	0.525	Wightman 1981
Acetic Acid	O_2	flow	$2.55 \bullet 10^{11}$	172.7	1	0	611-718	394-438	0.525	Wightman 1981
Activated Sludge (COD)	O_2	batch	$\sim 1.5 \bullet 10^2$	~54	1	0	573-723	240-350	46.5	Shanableh 1990
Ammonia	O_2	flow	$3.16 \bullet 10^6$	157	1	0	913-973	246	0.03-0.11	Webley et al. 1991
2-Butanone	O_2	batch	$1.20 \bullet 10$	36.2	1	0	673-773	240-400	~6	Griffith and Gloyna 1992
Carbon Monoxide	O_2	flow	$3.16 \bullet 10^6$	112	1	0	673-814	246	0.02-0.11	Helling and Tester 1987
Carbon Monoxide**	O_2	flow	$3.16 \bullet 10^8$	134	0.96	0.34	693-844	246	0.01-0.098	Holgate et al. 1992
o-Cresol	O_2	batch	$3.16 \bullet 10^0$	28.5	1	0	673-773	240-400	~10	Griffith and Gloyna 1992
Digested Sludge (COD)	O_2	batch	$4.36 \bullet 10^3$	20.4	1.86	0	573-723	240-350	46.5	Tongdhamachart 1991

2,4-Dichlorophenol	O_2	flow	$1.94 \bullet 10^{4}$	71.9	1	0.38	683-788	276	0.4-0.8	Crain and Gloyna 1992
Ethanol	O_2	flow	$6.46 \bullet 10^{21}$	340	1	0	755-814	241	0.03-0.036	Helling 1986
Formic Acid	O_2	flow	not reported	~96	1	1	683-691	408-432	1.0	Wightman 1981
Glucose (TOC)	O_2	batch	not reported	130	0.5	1	653-683	~400	~10	Whitlock 1978
Methane	O_2	flow	$1.26 \bullet 10^{7}$	156.8	1	0	913-973	245	–	Rofer and Streit 1989
Methane	O_2	flow	$2.51 \bullet 10^{11}$	178.9	0.99	0.66	833-903	245	–	Webley, Tester, and Holgate 1991
Methane	O_2	flow	$2.04 \bullet 10^{7}$	141.7	1	0	833-903	245	–	Webley, Tester, and Holgate 1991
Methanol	O_2	flow	$2.51 \bullet 10^{24}$	395.0	1	0	723-823	243	–	Rofer and Streit 1989
Methanol	O_2	flow	$3.16 \bullet 10^{26}$	408.4	1.1	-0.02	723-823	243	0.038-0.17	Webley et al. 1990
Phenol	O_2	flow	$2.61 \bullet 10^{5}$	63.8	1	1	557-702	292-340	0.1-0.4	Wightman 1981
Phenol	O_2	flow	–	–	0.5	0	653	188-278	0.25-1.0	Thornton and Savage 1990
Pyridine	O_2	flow	$3.44 \bullet 10^{14}$	227	1	0.2	698-800	276	1-3	Crain and Gloyna 1992

*Kinetic parameters are defined by $-d[C]/dt = k[C]^m[O]^n$ and $k = k^{\bullet}\exp(-E_a/RT)$, where [C] and [O] are concentrations of organic reactants and oxidant, respectively; E_a is in kJ/mol; T is in K; R = 8.314 J/mol-K; and $k^{\bullet}$ = 1/sec (first-order), etc. – Not available. $[C]_o$ = feed concentration. The concentration of compounds labeled with COD is quantified by chemical oxygen demand method; the concentration of other compounds is quantified by chromatographic techniques. The excess oxidants are used in all tests. Kinetic parameters are reported for the overall reaction in water unless otherwise indicated.

**Parameters have been obtained for oxidation only (e.g., excluding reactions with water).

Reprinted from *Proceedings of the Second International Symposium on Environmental Applications of Advanced Oxidation Technologies* , Gloyna and Li, "Supercritical Water Oxidation: An Engineering Update," 1993 with permission of EPRI.

Three general chemical reaction kinetics principles apply in a SCWO system. First, the oxidation rate is independent or only weakly dependent on the oxidant concentration. Therefore, only a small excess (over the stoichiometrically required) amount of oxygen needs to be fed to the SCWO reactor to result in very high oxygen utilization. Second, the oxidation reactions generally follow pseudo first-order kinetics with respect to the concentration of starting compounds. Third, the activation energy for organic compounds treated in a SCWO system ranges from about 20 kJ/mol (4.8 kCal/mol) to 408 kJ/mol (98 kCal/mol). As a result, the residence times required for a desired level of treatment can be determined by applying batch reactor treatability study results to a first-order reaction model for a plug flow reactor.

Figure 4.2
Generalized Idealized Regimes for SCWO Reactor Operations

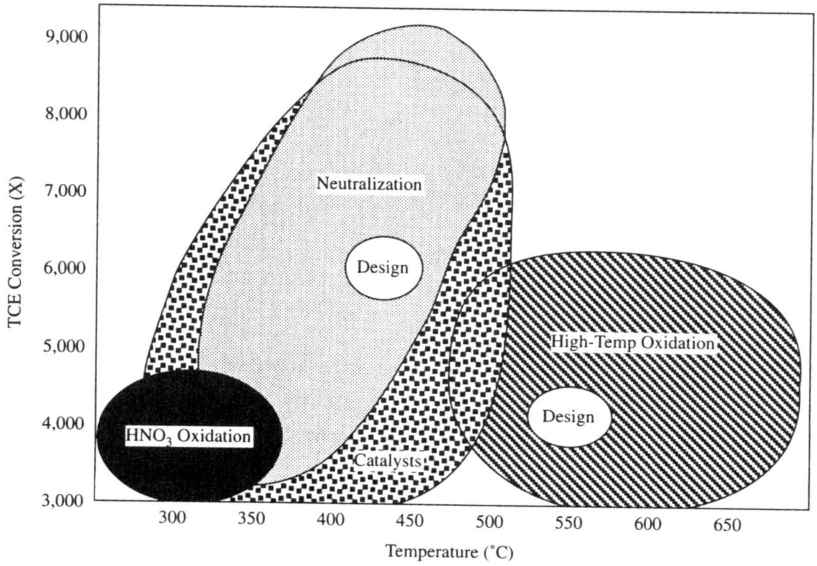

Reprinted from *Proceedings of the Second International Symposium on Environmental Applications of Advanced Oxidation Technologies*, Gloyna and Li, "Supercritical Water Oxidation: An Engineering Update," 1993 with permission of EPRI.

4.5.2 Design and Equipment Selection

The first step in the design of a SCWO system is a series of treatability tests to establish the reactor's operating regime. Figure 4.2 illustrates the various conceptual regimes which have been proposed for SCWO reactor operations (Gloyna and Li 1995).

The next step in the design of a SCWO system is specification of the size and shape of the reactor. The reactor's size is based on the required fluid residence time determined from treatability results and the desired waste throughput. Since the chemical reactions are approximately first order and the rates are independent of the oxygen concentration, testing of relatively few concentrations of organics can establish the required reactor volume for the "worst-case" conditions. Using the equations describing first-order reaction kinetics (Levenspiel 1962) allows for sizing of the reactor over a wide range of concentrations from tests at only two conditions; however, it is recommended that the treatability tests include several additional contaminant concentrations to verify that first-order kinetics apply to the specific case.

Once the reactor's volume has been determined, it is necessary to select its diameter and length. The diameter is set to maintain the fluid velocity above a minimum value for tubular reactors (determined through treatability studies on the specific waste streams) that minimizes deposition. During treatment, inorganic compounds will precipitate and can clog reactor components. It is not possible to predict the particle size of the precipitate from first principles, so precipitation and deposition studies for the waste to be treated should be conducted as part of the treatability study. These tests should be conducted on a pilot-scale system to allow for realistic liquid velocities and fluid dynamics.

The next design determination is the amount of heat that must be supplied to the process. The use of external heating reduces the amount of oxygen that must be used and compressed into the system. External heat can be supplied by gas or oil burners or by electric heaters. External gas or oil heating results in a lower oxygen requirement, but increases the amount of fuel needed because of combustion and heat transfer inefficiencies. Direct injection of the fuel results in essentially complete utilization of its heating value, but requires that oxygen (or another oxidizer) be supplied and energy be expended pumping the oxygen or oxidizer into the high-pressure system. The tradeoff must be evaluated on a case-by-case basis. The quantity of heat that is needed will also depend on the efficiency of the heat recovery system.

Again, a site-specific evaluation of the tradeoff between the cost of the heat exchange equipment and the cost of fuel is required.

4.5.2.1 Materials of Construction/Corrosion Management

For most SCWO applications, the heat-exchange equipment and the reactor will have to be made of specialty alloys and some parts may need to be lined with materials such as platinum or titanium for corrosion protection. In some applications, special corrosion resistant materials of construction may not be required for transpiring wall reactors, discussed below. As a result, the increased cost of the corrosion-resistant heat recovery system must be carefully weighed against the cost savings from reduced fuel usage. This analysis is further complicated by the fact that increased fuel usage can result in either a greater volume for the reactor (if the fuel is injected) or a greater heat transfer area if external heating is used. Such increased size requirements increase the need for high-priced materials of construction and, therefore, increase system capital cost.

The point at which the liquid effluent, which contain the inorganic acids, cool below the critical point typically is also the point where aggressive corrosion attack is most likely. Various means have been employed to minimize this attack. The most obvious is the use of materials or liners which are resistant to this type of corrosion. Data on the corrosion resistance of a range of materials is presented elsewhere in this chapter. An alternative means of protection is the use of sacrifical liners. These are attacked preferentially and replaced on a regular schedule. The electrochemical potentials thus set up protect the lines and fittings. An alternative method for protecting the lines is employed in the Aerojet transpiring wall reactor whereby a non-corrosive clean fluid is injected into critical parts of the system in such a way as it forms a non-corrosive boundary layer at these corrosion-prone points. Clearly, the type of protection used will be determined by the nature of the material to be treated and the relative costs of the alternatives.

Specification of the SCWO reactor and ancillary high-pressure/high-temperature equipment must address two materials problems: (1) corrosion and (2) high pressures at high temperatures. Under these conditions, metal alloys tend to embrittle and experience creep. This material degradation, coupled with the possibility of corrosion-induced pitting, cracking, or crazing, creates a potential design problem (Blaney et al. 1995). In some cases, passive corrosion control will require the use of some form of lined or composite materials of construction. Reactors which use a dynamic means of corrosion control, such as the transpiring-wall platelet reactor, described below, may require less expensive materials of construction. For example, a passive design developed by Kimberly-Clark (patent pending) consists of a reactor surrounded by a carbon steel pressure vessel which is maintained at a substantially lower temperature than the reactor. The reactor is designed to withstand the high temperature and corrosive environment, but is incapable of withstanding the full pressure. The surrounding vessel is strong enough to withstand the pressure. The space between the reactor and the outer vessel is filled with an insulating fluid. Control systems maintain the pressure of the insulating fluid at approximately the same pressure as that of the reactor, resulting in lower pressure stresses on the temperature- and corrosion-resistant reactor.

General Atomics conducted corrosion testing in developing a SCWO system to treat propellants and chemical warfare agents. The results of this testing are summarized in Table 4.5. Platinum was identified as the most chemically resistant material of construction for SCWO processing of GB and VX agents. Therefore, a platinum-lined Hastelloy C276 reactor was used for GB and VX treatment. For processing the hydrolysates of mustard agent or solid propellant, titanium was identified as the best material of construction, again as a thin liner within a Hastelloy C276 pressure-bearing wall (Hazlebeck, Downey, and Roberts 1994; Turner 1993). Both of these applications required relatively high temperatures and pressures to achieve the desired endpoints. Because treatment conditions required for many remediation projects are not as rigorous, less exotic materials of construction may be acceptable.

Corrosion testing conducted in a MODAR vessel reactor (INEL 1995) has shown that proper choice of materials of construction can allow operation of a SCWO system with corrosion maintained at an acceptable level.

A new type of reactor design that was recently developed attempts to overcome corrosion and deposition problems in a dynamic manner. The reactor, developed by GenCorp Aerojet, is termed a *platelet liner* technology. It has been successfully tested by Sandia National Laboratories and is being incorporated by Foster Wheeler Development Corporation into SCWO demonstration units for destruction of certain smokes and dyes for the U.S. Army and of shipboard hazardous materials for the Navy.

Table 4.5
Corrosion Results Summary

	HF and H_3PO_4 (GB)			H_2SO_4 and H_3PO_4 (VX)			HCl and H_2SO_4 (Mustard)		
Material	350°C	450°C	550°C	350°C	450°C	550°C	350°C	450°C	550°C
Pt	□	□	□	□	□	□	◇	△	□
Pt/Ir	□	□	□	□	□	□	◇	△	□
Pt/Rh	□	□	□	□	□	□	◇	△	□
Hf	◇	◇	◇	△	△	△	◇	◇	◇
Ti	△	◇	△	□	◇	△	△	△	△
Timet 21S	□	◇	◇	□	◇	△	△	△	△
Zr 704	◇	◇	◇	△	△	△	◇	◇	N/A
Mo	◇	◇	◇	◇	◇	◇	◇	N/A	◇
Nb	△	△	◇	△	◇	◇	◇	◇	N/A
Nb/Ti	△	◇	◇	□	◇	△	△	△	△
Ta	□	◇	◇	□	□	◇	□	◇	N/A
Al_2O_3	◇	◇	◇	△	◇	◇	◇	◇	△
AlN	◇	◇	◇	□	◇	△	◇	◇	◇
Sapphire	◇	◇	◇	△	△	△	◇	◇	△
Si_3N_4	◇	◇	◇	◇	◇	◇	◇	◇	◇
SiC	◇	◇	◇	◇	◇	◇	◇	◇	◇
ZrO_2	△	◇	◇	△	◇	△	◇	◇	△
C22	□	◇	◇	△	◇	◇	◇	◇	◇
Hast. C276	△	◇	◇	△	◇	◇	◇	◇	◇
Hayn. 188	△	◇	◇	△	◇	◇	◇	△	◇
HR-160	△	◇	◇	◇	◇	◇	◇	◇	◇
Inc. 825	□	◇	◇	△	◇	◇	◇	◇	◇
Inc. 625	□	◇	◇	△	◇	◇	◇	△	◇

□ Good (<10 mi/yr corrosion rate)
△ Moderate (10-200 mi/yr corrosion rate)
◇ Poor (>200 mi/yr corrosion rate)
N/A Not available

Reproduced courtesy of General Atomics

Figure 4.3
Transpiring-Wall Platelet Reactor

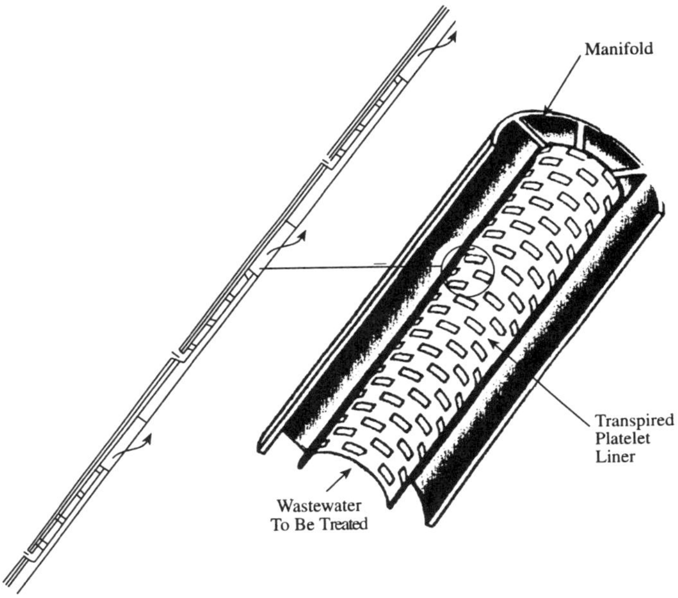

Reproduced from Rousar, Young, and Sieger, *Development of Components for Waste Management Systems Using Aerospace Technology*, 1995 courtesy of Aerojet Corp.

The transpiring wall platelet reactor consists of an outer cylindrical pressure housing, which encloses a concentric cylindrical platelet liner with a small, but finite gap between the two (Figure 4.3). Clean high-pressure water enters the pressure housing through inlet nozzles and feeds into inlet manifolds located strategically on the platelet liner outer surface. Following an intricate circuitry through several platelets, the high-pressure water stream is metered, split repeatedly, and delivered to the liner inner surface through numerous injection points. The injected water forms a nonreactive barrier between the platelet liner and the reactants and reaction products (Rousar, Young, and Sieger 1995).

The transpiring-wall reactor offers additional advantages:

- The temperature of the platelet liner and the pressure housing is controlled by the transpiration water. This isolates the reactor from any high-temperature swings in the reaction. Furthermore, the reactor allows the option of operation at higher reaction temperatures which dramatically improves destruction efficiency and throughput.
- The reactor configuration allows design optimization by injection of transpiration water at different temperatures and flow rates along the length by manifolding the outer housing.

Long-term performance data on this type of reactor are now being collected using a pilot-scale system where throughput is 30 cc/sec (28.5 gal/hr). The costs and potential failure modes for such a dynamic system versus traditional, static methods of corrosion control must be considered in the overall determination of which design is to be used for a specific application.

4.5.2.2 Heat Transfer

Heat transfer is another important consideration. Water exhibits marked changes in heat transfer properties near its critical point (Michna 1990). Limited heat transfer data for supercritical water under turbulent-flow conditions (mass velocities ranging from 75 to 4,000 kg/m^2-sec) have been reported (McAdams, Kennel, and Addoms 1950; Dickinson and Welch 1958; Yamagata et al. 1972). A University of Texas at Austin study focused on heat transfer to supercritical water under laminar- to transient-flow conditions (mass velocities ranging from 2.6 to 49 kg/m^2-sec)(Michna 1990). Heat transfer to water was enhanced for bulk temperatures just below the critical point. Increased natural convection effects due to the extremely low kinematic viscosity of water near the critical point are believed to be partly responsible for such enhancement.

The critical point of a solution is similar in concept to the critical point of a pure compound. Each pure compound has a unique critical point defined by the temperature, pressure, and density (T-V-ρ), although from the "phase rule," if any two of these variables are specified, the third is fixed. For a solution, the critical T-V-ρ is not usually a unique point. Rather, it is a range of values for each variable within which the solution exhibits properties similar to those of a pure compound at its critical point. The T-V-ρ above which the solution exhibits the critical properties needed for a given application is commonly termed the critical point of the solution, but the fact that it is actually a range should be kept in mind. The critical temperature is the apparent critical temperature of the mixed materials, that is, the temperature and pressure at which the liquid and gas properties of the mixture approach one another; the critical temperature refers to the critical temperature of a pure substance.

Figure 4.4

Overall Heat Transfer Coefficient as a Function of Core Temperature

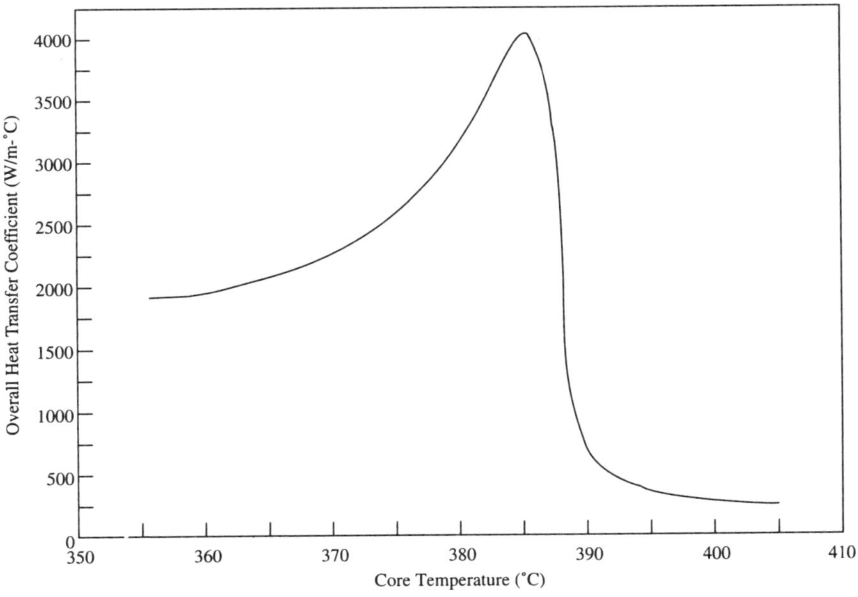

9.53 mm OD Tube
0.89 mm Wall
5.44 kg/hr
P = 248 bar

Reprinted from *Proceedings of the Second International Symposium on Environmental Applications of Advanced Oxidation Technologies*, Gloyna and Li, "Supercritical Water Oxidation: An Engineering Update," 1993 with permission of EPRI.

As shown in Figure 4.4, the heat transfer coefficient rapidly increases with increasing bulk temperature as the pseudo-critical temperature is approached. Deterioration in heat transfer occurs for bulk temperatures just above the critical temperature. Such deterioration appears to be largely the result of variations in the physical properties between the fluid at the surface of the tube and the bulk fluid. Therefore, deterioration is greatest for pressures close to the critical pressure where the physical properties change most rapidly.

Design correlations developed for high-temperature/high-pressure steam boilers and turbines can be used in the design of SCWO systems. Figure 4.5, which shows the viscosity of water/steam in the critical and supercritical region, illustrates the rapid changes in physical properties associated with this region. As can be seen, as the critical temperature is approached (374.1°C [705.4°F], and 2.210 MPa [3,206 psi]) the fluid's viscosity drops rapidly, approaching that of a gas rather than of a liquid.

The operating conditions for the SCWO process can be modified to take advantage of the changes in the solubility of inorganic materials at different temperatures and pressures so as to remove inorganic contaminants as a concentrated stream. For example, at high temperature (800°C [1,472°F]) and relatively low pressure (240 bar [3,500 psi]) most sticky salts will precipitate and then melt, thereby, tending to adhere to the walls of the system. Metal oxides, on the other hand, tend to have higher melting points and therefore, do not adhere to system surfaces. Most salts will precipitate at pressures of about 242 bar (3,500 psi) and temperatures of about 400°C (752°F). However, at a higher pressure (1 kbar or 14,570 psi) and a lower temperature (500°C [932°F]) and most of the salts will remain in solution. The higher pressure (1 kbar) also causes a relatively large increase in water density (0.3 to 0.8 g/mL) without an accompanying increase in viscosity (approximately 0.05 cP).

A variety of methods have been developed (such as filters or hydrocyclones) to remove a large fraction of the inorganic solids (salts and inorganic oxides) from the effluent. Salts are generally sticky under SCWO conditions and, hence, they are more difficult to remove and tend to clog the system. Metal oxides are generally not sticky and, hence, are easier to remove and cause fewer clogging problems than salts.

One means of reducing the impact of sticky salts on a system is to reduce the number of surfaces on which they can deposit and lead to plugging. The transpiring-wall reactor described above is one solution being proposed to minimize the deposition of sticky solids on the reactor walls; however, it

Figure 4.5

Viscosity of Water and Water Vapor in the Critical Region

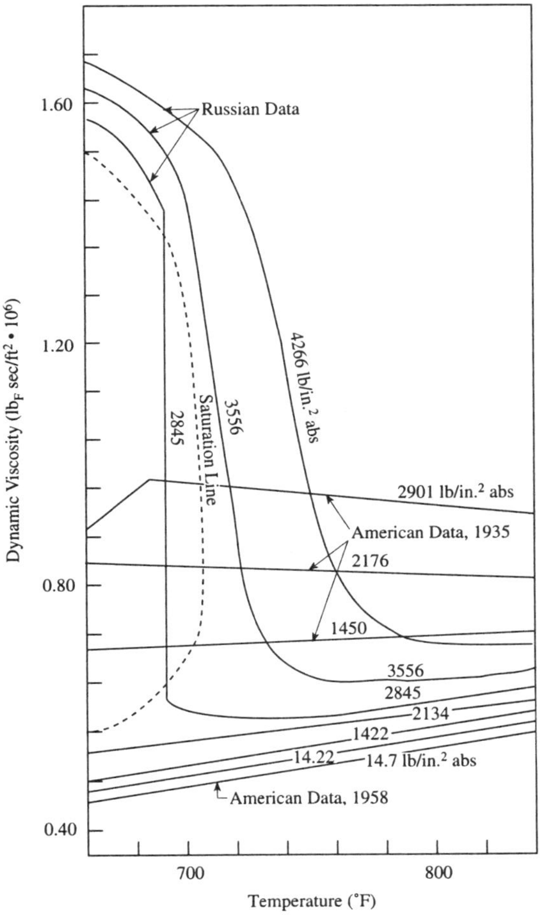

Reproduced with permission of The McGraw-Hill Companies from Perry, Chilton, and Kirkpatrick, *Chemical Engineers Handbook*, 1993.

does not eliminate these problems entirely. Deposition and plugging can still occur at the inlet and outlets to the transpiring-wall reactor as well as on the surfaces of ancillary heat transfer equipment. Finally, some salts can redissolve when the effluent from the SCWO reactor is cooled and depressurized to below the critical point of water. Impact of these salts on the effluent quality must be considered. Development is continuing on the use of filtration, hydrocyclones, and flushing methods and vendors should provide data validating their proposed method of dealing with solids deposition for each particular application.

Even though they are not sticky, oxide particulates can also clog narrow passages, inlets, and outlets of the reactor; however, these are more easily removed from the system than sticky salts by means such as settling, filtration, or hydrocyclones. The performance of small hydrocyclones for non-sticky solids is illustrated with data derived from a UT pilot plant (Dell'Orco 1991; Dell'Orco 1993, Dell'Orco et al. 1991a). Mln-U-Sil 5 (quartz silica, U.S. Silica Corp.), exhibiting density and particle-size characteristics similar to the oxides formed in reactors during SCWO operations, is typical of various particles evaluated. To evaluate hydrocyclone performance, it is necessary to determine solids separation efficiency and particle-size distributions. Equation 4.5 can be used to estimate solids separation efficiency as a function of Stokes' Number (W):

$$E_G = 1 - A\Psi^B \tag{4.5}$$

where: Ψ = $\{(\rho_p - \rho_w) \bullet d^2 \bullet v_i\} / 18\mu D$

ρ_p, ρ_w = density of particle and water respectively, kg/m^3

D = diameter of cyclone at feed port, m

v_i = fluid inlet velocity, m/sec

d = characteristic particle diameter, m

A,B = empirical constants (0.018 and 0.64, respectively, for 10 mm (0.4 in.) cyclone and silica, titania, and zirconia powder, to be determined by treatability testing for specific application; and

E_G = gross separation efficiency (dimensionless).

Figure 4.6 shows experimental gross separation efficiency data for two hydrocyclones (10 mm and 25.4 mm diameters [0.4 and 1 in.]) and the model calculation for the 10 mm (0.4 in.) hydrocyclone. Gross separation efficiencies near 80% are achievable for silica at temperatures above

Figure 4.6

Gross Separation Efficiency (as penetration) for Two Hydrocyclones
(Hydrocyclone A: 10 mm diameter; Hydrocyclone B: 25.1 mm diameter)

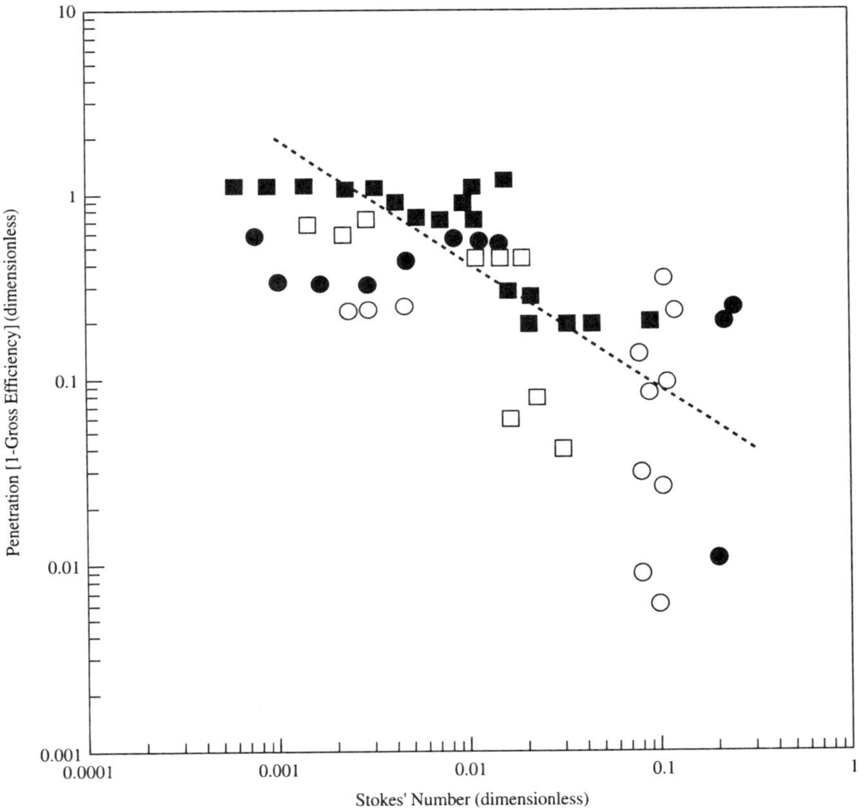

□ Hydrocyclone A, Titania
○ Hydrocyclone A, Zirconia
■ Hydrocyclone A, Silica
● Hydrocyclone B, Silica
- - - Data Fit, Hydrocyclone A

Reprinted from *Proceedings of the Second International Symposium on Environmental Applications of Advanced Oxidation Technologies*, Gloyna and Li, "Supercritical Water Oxidation: An Engineering Update," 1993 with permission of EPRI.

300°C (572° F), while at the same temperature range, gross separation efficiencies for the more dense zirconia particles are greater than 99%. The individual silica particle-size (grade) separation efficiencies are shown in Figure 4.7. The separation efficiency of these solids is directly tied to particle size, temperature, and pressure, although particle size would be less of a problem if filtration were employed instead of a cyclonic separator. The objective is to accurately predict separation efficiencies for individual particle diameters. Equation 4.6 can be used to predict separation efficiencies of bulk solid streams and individual particle sizes (Leith and Licht 1972):

$$h = 1 - \exp\{-2C\Psi(n+1)^{\{1/(2n+2)\}} \tag{4.6}$$

where:
- h = separation efficiency;
- C = geometrical parameter (dimensionless);
- n = vortex exponent (dimensionless); and
- Ψ = as defined for Equation 4.5.

Chromium speciation during SCWO is of considerable importance (Rollans, Li, and Gloyna 1992). While the primary concern is the chromium in the waste streams being treated, a secondary concern is the presence of chromium in many "stainless" steels that might be used for the reactor or other components and could leach due to the corrosive environment. Chromium behavior was studied in a bench-scale, vertical, concentric-tube reactor system treating municipal wastewater sludges to determine the hexavalent and soluble trivalent chromium concentrations in the reactor bottom and the treated effluent. The reactor material was Stainless Steel 316.

Under SCWO conditions, hexavalent and trivalent chromium corrosion products were generated and removed by precipitation. At 400°C (752°F) and a Reynolds number of approximately 8,000, the hexavalent and trivalent chromium concentrations in the effluent were <0.004 mg/L and 0.163 mg/L, respectively. Similarly, with an influent feed of 8.76 kg (19 lb) of sludge, the concentrations of hexavalent and trivalent chromium in the bottom of the reactor were 0.288 mg/L and 3.712 mg/L, respectively.

Optimum separation efficiency, species distribution, and corrosion effects depend on the type of sludge and treatment conditions. Two mechanisms appear to be responsible for separation of hexavalent and trivalent chromium from effluents: (1) the precipitation of chromate complexes and trivalent chromium salts, and (2) the settling of solid residues, including sorbed trivalent chromium. In the case of the municipal sludge, hexavalent chromium, in the form of insoluble chromate complexes, settled from the bulk supercritical fluid. The concentration of the settled hexavalent chromium in the reactor bottom was at least 72 times greater than the chromium concentration in the effluent. At the subcritical and laminar-flow conditions, the concentrations of chromate complexes in the effluent and reactor bottoms were 0.046 mg/L and 0.035 mg/L, respectively.

Figure 4.7

Grade Efficiency for Hydrocyclone Separation of Silica

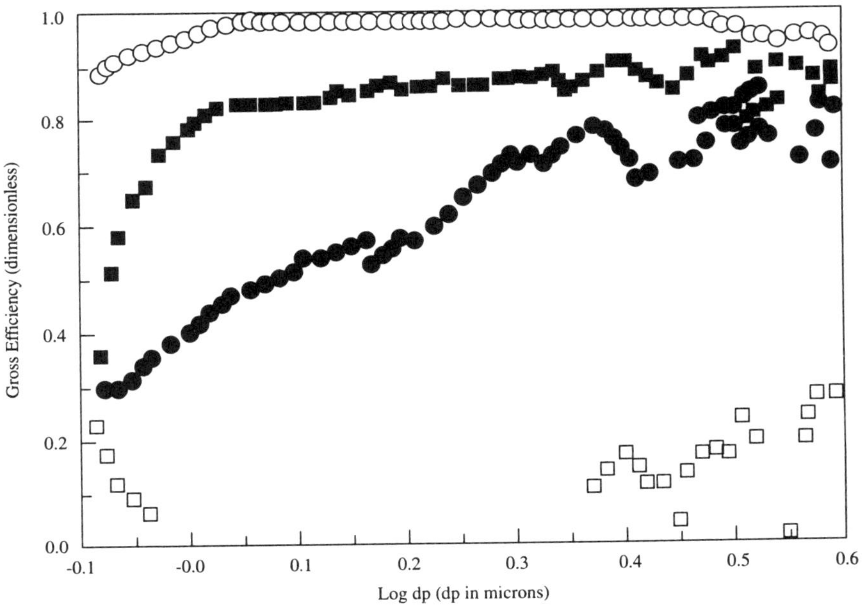

□ 70°C
● 147°C
■ 213°C
○ 371°C

Reprinted from *Proceedings of the Second International Symposium on Environmental Applications of Advanced Oxidation Technologies*, Gloyna and Li, "Supercritical Water Oxidation: An Engineering Update," 1993 with permission of EPRI.

For both temperature regimes used during the treatment of the industrial sludge, the concentration of hexavalent chromium was below the detection limit in the effluent and reactor bottom. A large portion of trivalent chromium was sorbed onto the solid residues which settled in the reactor bottom. Concentrations of soluble trivalent chromium in the effluent and reactor bottom were comparable.

In general, moderate amounts of chromium can be tolerated in the reactor's materials of construction. The possibility of trivalent chromium being converted to hexavalent chromium appears small. Hexavalent chromium present in the influent appears to precipitate out of solutions at typical SCWO reactor conditions and, if chromium is an effluent problem, it appears possible to incorporate particulate removal into the system and collect the chromium as a separate concentrated stream.

4.5.3 Process Modification

The SCWO process is highly flexible and can be modified to accommodate a wide variety of conditions and waste streams. Modifications that have been proposed include:

- use of deep wells to reduce pumping costs and the pressure drop across the reactor lining; and
- use of catalysts to increase the rates of chemical reaction.

Deep well reactors are an adaptation that have been used for WAO systems, but apparently have never been applied to SCWO systems where the much higher temperatures and pressures make it difficult, if not impossible, to implement. Conceptually, the approach is simple. A well is drilled into the earth (a stable and impervious geologic formation is a requisite) down to a sufficiently deep level so that the hydrostatic pressure equals the reactor's operating pressure. To illustrate, the well would have to be approximately 2,252 m (7,400 ft) deep to achieve the critical pressure of water at its bottom. Then, the well is lined, and a smaller annular pipe is dropped to within a foot or two of the bottom. The annular pipe is braced against the outer casing in such a manner that liquid will freely flow past the bracing. Heaters and pipes feeding pressurized oxygen or another oxidizer would also need to be lowered into the lower part of the well.

The contaminated liquid is pumped down into the well where it is heated as the liquid's hydrostatic pressure increases. The combination of hydrostatic pressure and heat generated by the oxidation of the organic compounds in the waste turns the volume at the bottom of the well into a SCWO reactor. The cost of pumping is reduced and, because the reactor is in the well, the pressure drop across its lining is lessened. The pressure stresses on the reactor are similar to those of the pressure stresses across the inner reactor in the Kimberly-Clark reactor described in Section 4.5.2. While the approach is intriguing, the difficulties associated with maintaining a tight seal and maintenance of the system 2.2 kilometers (almost one and one half mile) underground diminishes its viability at the current state of development.

Catalysts can enhance the total conversion of complex organo-nitrogen compounds (including lower molecular weight transformation products), shorten the reaction time, and lower the required reaction temperature. However, the presence of inorganic materials can result in depositions on the catalysts and reduce catalyst life. The use of catalysts has been tested at a laboratory-scale on WAO systems to good effect and a large body of research data on the subject is available (Gloyna and Li 1993).

4.5.4 Pretreatment Processes

The only physical pretreatment that may be required for SCWO is some form of screening followed by masceration to remove or destroy solids too large to pass through the pump. Generally, abrasive solids need to be removed or size reduced sufficiently to protect the high-pressure pumps, valves, and seals.

A common form of pretreatment used for many highly-reactive organic compounds such as chemical agents or rocket fuels is hydrolysis. During hydrolysis, the waste is mixed with water, often containing a caustic. Hydrolysis reduces the wastes reactivity and, in the case of many chemical munitions agents, toxicity to the point where they are safer to handle. Sufficient water is added to bring the wastes' heating value to within that required for the SCWO reactor. SCWO is currently under active evaluation as a means of treating such hydrolyzed highly-reactive wastes.

4.5.5 Posttreatment Processes

The SCWO system produces three by-product streams: (1) aqueous product, (2) solid precipitate or filtrate, and (3) gas. The aqueous product consists of the water and dissolved solids and residual organic compounds, if any, that are products of the SCWO chemical reactions. The solid precipitate consists of oxides and other insoluble inorganic materials, and the gas stream consists of nitrogen (if air is used as an oxidizer), excess oxygen, carbon dioxide, and water vapor with possible traces of carbon monoxide, sulfur oxides, nitrogen oxides, and organic constituents.

The aqueous product coming from a properly designed and operated SCWO system contains extremely low levels of organic material, often below the levels of detectability. Low concentrations of acetic acid or ammonia are found. These materials can be removed to below detection limits by increasing the temperature or the residence time of the SCWO reactor. Alternatively, they can be removed by subjecting the effluent to normal biological treatment in a wastewater treatment plant. The amounts of hazardous products of reaction (the equivalent of products of incomplete combustion, PICs, in an incinerator) that occur in the SCWO effluent, and indeed in all the waste streams from a properly specified and operated SCWO system, are extremely small. The case study and results presented in this chapter indicate the levels that have been encountered during testing under various conditions.

Inorganic materials, such as metals, will pass through the SCWO system or be deposited in it, the latter being undesirable. SCWO conditions oxidize most metals and these will precipitate. In some applications, where the waste being treated has a high metals content, some form of precipitation or ion exchange may be needed.

The solid wastes from a SCWO system are the precipitates, filter cakes, and internal system deposits which are removed during maintenance. Their composition is completely site-specific and no general posttreatment methods can be recommended.

The small-volume gas stream emanating from the SCWO system consists largely of the following constituents:

- excess oxygen pumped into the process;
- water vapor;
- nitrogen from any air that may have been pumped into the reactor as oxidizer and from nitrogen-bearing constituents in the waste; and
- carbon dioxide which is the product of oxidation of the organic compounds in the waste.

In addition to these relatively major constituents, trace gases may be found in very low concentrations in the emission stream from a SCWO system. The concentrations found are usually much lower than those found in emissions from any type of combustion equipment. More important, when oxygen (rather than air) is the oxydent, gas emissions are much smaller than from combustion systems. The following materials might also be present in the emissions:

- CO_2, traces of CO sometimes occur as normal equilibrium products of the oxidation reaction;
- ammonia, if nitrogen is present (depends on reactor conditions);
- oxides of sulfur, if sulfur is present in the waste (depend on reactor conditions);
- acetic acid.

While traces of PICs such as chlorodibenzodioxins and chlorodibenzofurans (dioxins, furans) have been found in the effluent from a SCWO reactor operated at lower temperatures (see Section 4.7.3), these compounds have a very low volatility and are not carried into the gas stream in a SCWO reactor. It is noted that higher operating temperatures and pressures can destroy these contaminants directly in the SCWO reactor.

The particulate concentration in the gas stream is negligible and because of the very low flow rates the stream can be readily treated by standard air pollution control techniques, such as adsorption, to control the other trace contaminants. In most cases, such treatment will not be required.

4.5.6 Process Instrumentation and Controls

The process requires careful control to maintain the required temperature and pressure of the reactor and ancillary units. Because of the large changes in physical properties and inorganic compound solubilities that occur in the super-critical region, this control has relatively small tolerance. The process is controlled through the flow rate at the waste pump, the water pump, the oxidant system, and the temperature in the reaction zone. Feedback is received through pH and TOC analysis of the liquid effluent. Other than these special requirements, the instrumentation and control needs of the system are equivalent to those of a very small hydrocracker which is common in the petroleum industry.

4.5.7 Safety Requirements

A SCWO system operates at relatively high pressures — typically on the order of 276 bar (4,000 psi). This fact is sometimes pointed out as a safety concern. While this is an understandable assessment, a more careful analysis of a SCWO system reveals that the risk is less than or equivalent to those of common industrial processes, primarily because of the minimal amount of energy in the high-pressure portion of the system.

The SCWO system requires routine safety systems and temperature and pressure controls as well as routine safety procedures associated with the industrial use of pressurized liquid and oxygen. The safety requirements are analogous to those of very small capacity industrial processes used in the petroleum refining industry.

4.5.8 Specification Development

The specifications for a SCWO system include the following major items:

- **Quantity of material to be treated and rate of treatment.**

  This information is required to size the system and to establish the system's anticipated life. The anticipated life is needed to both establish a cost and to specify materials and a maintenance and parts replacement schedule so that the system will withstand the harsh environment for its anticipated life.

- **Total organic content of the waste stream.**

  This information is needed to determine the amount of dilution, heat recovery, external heat, added fuel that is needed to achieve the desired temperatures, and residence times.

- **Maximum quantities of organic halogens, phosphorus, and sulfur.**

  This information is used to establish the amount of neutralizing chemicals that will be required, to select corrosion-resistant materials, solids separation processes needed, and operational programs required to achieve the desired treatment.

- **Specific contaminants that must be destroyed and acceptable minimum levels of destruction.**

  This information is needed to establish the reactor residence time and volume.

- **Quantity and size of suspended solids in the feed material.**

  This information is needed to determine size reduction and pumping requirements.

- **The types and quantities of soluble inorganic constituents that are present.**

  This information is needed to establish the type and amount of insoluble material that is present at the reactor conditions and to ascertain the "stickiness" of the precipitates, where in the system they may come out, and whether these solids pose a deposition or clogging problem during operation. For example, if non-sticky solids such as metal oxides are found to precipitate from the supercritical fluid in the reactor, a tubular reactor must be designed so that the fluid flow rates are high enough to sweep them out of the system. For a transpiring-wall reactor, these inorganic concentrations will be used to establish the required water flow rates through the liner.

  Sticky solids such as mineral salts also precipitate in the supercritical fluid and these may adhere to the reactor wall. Various proprietary methods have been developed by different SCWO vendors to minimize or avoid buildup of these salts on system surfaces. The transpiring-wall reactor is one example of the different approaches to this design challenge.

- **Discharge requirements for organic and inorganic constituents.**

  This information will establish the temperature and residence time required to achieve the desired level of organic compound destruction, and whether posttreatment systems such as filtration or ion exchange systems are needed to remove suspended or dissolved inorganic materials.

- **Available utilities at the site.**

4.5.9 Cost Data

Cost analyses of a SCWO system for treating paper mill sludge were performed independently by Kimberly-Clark Corporation and by Charles Eckert of the Georgia Institute of Technology (Blaney et al. 1995). The details of the system and of the treatment parameters are given in the case study in Section 4.7. The results were comparable and indicated a cost of \$33 to \$44/wet tonne (\$30 to \$40/wet ton) of sludge (dewatered to 50% solids). These costs included a credit for recovered calcium carbonate ash from the SCWO unit.

There is no technical reason why SCWO cannot be applied to wastewaters containing a wide range of organic material concentrations. However, below concentrations which are autogenic (i.e., the organics in the ground or surface waters being treated provide sufficient energy for heating and pumping it (and the oxygen source) required affects the process cost-effectiveness. Table 4.6 provides the steps of a simple cost analysis that can be used to determine the amount and approximate cost of external energy required for an SCWO unit. This analysis is a "worst-case" because to assumes no heat recovery and no energy released from the oxidation of the waste.

The calculations of Table 4.6 are for a 37.85 L/min (10 gal/min) system operating at 242 bar (3,500 psi) and 538°C (1,000°F). The calculation assumes that electricity is used to drive the compressors and that either electricity, natural gas, or fuel oil is used for external heat. External gas heating was assumed to have a 75% thermal efficiency, oil heating a 60% efficiency, and the overall pump/motor efficiency of 50%. Electricity is assumed to cost 6.0¢ per kWh, natural gas \$7.00 per MBTU, and fuel oil at 60¢ per gal. Other assumptions are shown in Table 4.6.

As this calculation shows, the cost of pumping the material up to the operating pressure of the reactor is only a minor component of the overall cost of the system, about 6.3¢ per 1,000 gal. The cost of heating can be significant ranging in cost per 1,000 gal from \$204.72 for electric heating to \$85.70 for oil heating. It is estimated that these costs can be cut in approximately half with appropriate use of heat recovery. The economics clearly favor SCWO for the treatment of wastes containing high concentrations of organic compounds whose energy content can be used towards heating the waste streams. Optimum economics for any applications requires a far more detailed analysis than this one which is only intended to illustrate the magnitude of the various costs.

Table 4.6
Power Cost Analysis for SCWO

Cost of Pumping to SCWO Operating Pressure		
Critical Pressure of Water	3,206.00 lb-ft/in.^{2}	From Perry's P 3-192
Critical Temperature of Water	705.4°F	From Perry's P 3-192
1. Pressure of Water	3,500.00 lb-ft/in.^{2}	
2. Assumed Flowrate	10 gal/min	
3. Conversion Factor	7.4805 ft^3/gal	
4. Assumed Flowrate	1.3368 [2/3] ft^3/min	
5. Conversion Factor	144 in.^{2}/ft^2	
6. Pressure of Water	504,000 [1 • 5] lb-ft/ft^2	
7. Pumping and Compressor Power	673,751.75 [8 • 6] ft-lb/min	
8. Pumping Power	11,229.20 ft-lb/sec	
9. Conversion Factor	0.00181818 horsepower/(ft-lb/sec)	
10. Pumping and Compressor Power	20.417 [9 • 8] horsepower	
11. Conversion Factor	0.001356 kW/(ft-lb/sec)	
12. No Loss Power	15.23 [10 • 11] kW	
13. Assumed Pump/Motor Efficiency	40.00%	
14. Power Usage for Pumping and Compressor	38.0670 [12/13] kW	
15. Conversion Factor	3,600 sec/hr	
16. Power Usage for Pumping and Compressor	0.00105742 [14/(2 • 15)] kWh/gal	
17. Assumed Cost/kWh	$0.060 $/kWh	
18. Pumping Power Cost	$0.00006344 [17 • 16] per gal	
19. Pumping Power Cost	$0.06344496 per 1,000 gal	
Cost of Heating to SCWO Operating Temperature		
Feed Water Temperature	60°F	
Water Enthalpy	28.06 Btu/lb @ 1,000°F	From Perry's P 3-191
Water Enthalpy	1,424.5 Btu/lb	From Perry's P 3-192
ΔH	1396.44 Btu/lb	
Conversion Factor	8.34 lb/gal @ 60°F	
Heat Requirement	11,646.31 Btu/gal	

Table 4.6 cont.
Power Cost Analysis for SCWO

Gas Heating			
Cost of Gas ($/MBtu)	$7.00		
Heat Cost per gal @ 100% Efficiency	$0.0815		
Assumed Heating Efficiency	75.00%		
Heating Cost per gal	$0.1087		
Electric Heating			
Cost of Power	$0.06		
Conversion Factor kWh/Btu	0.000292875		
Power per gal @ 100% Efficiency	3.4109		
Heating Cost per gal	$0.2047		
Oil Heating			
Cost of Oil, ($/gal)	$0.60		
Heating Value of Oil, Btu/gal	136,000		
Heat Cost per gal @ 100% Efficiency	$0.0514		
Assumed Heating Efficiency	60.00%		
Heating Cost per gal of Water	$0.0856		
Total Power Cost	**Electric**	**Gas**	**Oil**
per gal of Water	$0.2047	$0.1087	$0.0856
per 1,000 gal of Water	$204.72	$108.76	$85.69

4.5.10 Design Validation

The major design problems for SCWO systems have focused on finding suitable materials to withstand the high temperature, pressure, and corrosive environment and to manage inorganic materials within the reactor for continuous operation. These problems, while not completely solved, appear to be sufficiently overcome to allow design, construction, and operation of full-scale systems. For example, corrosion problems are being addressed through the use of special metals as in the General Atomics, EcoWaste Technologies, Inc., Foster-Wheeler, and ceramic cladding as in the Kimberly-Clark Reactor. Plugging problems are being addressed in tubular reactors by maintaining the fluid flow high enough to scour the surfaces as, for example, in General Atomics reactors or by the use of unique designs which prevent the fluid being treated from coming in contact with the walls, as in the Aerojet transpiring-wall reactor.

SCWO appears appropriate to those situations where (1) the waste stream contains a sufficient organics content to provide the majority of the heating value needed for the high temperatures, or (2) the contaminants are so highly toxic that safe treatment demands a fully-enclosed process. SCWO is becoming increasingly competitive with incineration for hazardous waste disposal.

4.5.11 Permitting Requirements

Because it is a fully-enclosed technology, a SCWO system requires far fewer permits than, for example, an incinerator. If the system is treating a hazardous waste (as defined by the Resource Conservation and Recovery Act [RCRA]), then the requisite RCRA permits would be required. In addition, the system will require discharge permits for the liquid effluent and air permits for the small gaseous emissions. None of these permits should pose significant difficulties since a SCWO system would not be considered a major source of contaminants.

4.5.12 Performance Measures

There are no unique or unusual performance measures for SCWO systems. Performance measures are the percentage of organic content destroyed, length of time of continuous operation, mean time between failures (operational reliability), reactor life cycle, and the ability for the system to react to a change in the waste stream characteristics.

4.5.13 Design Checklist

Refer to the specification development guidance in Section 4.5.8.

4.6 Implementation and Operation

4.6.1 Implementation Strategies

Because of its present stage of development, implementation of SCWO requires a vendor experienced in the design, construction, and operation of such systems. The vendor must also have available a range of laboratory and pilot-scale equipment for conducting treatability studies. This equipment must be capable of achieving the range of operating conditions that encompass those required for the installation.

4.6.2 Start-up Procedures

Start-up procedures must be provided by the vendor for the specific equipment installed.

4.6.3 Operations Practices

Operating practices will be different for the different types of reactors and ancillary systems supplied by each vendor. The factors that must be considered to maintain optimum performance during operation of a SCWO system include the following:

- variability of the waste stream; and
- deterioration of the physical components of the reactor because of plugging or corrosion.

Variability of the waste stream flow rate should pose no particular problem for the system unless the flow exceeds the reactor maximum design throughput. Reduction in the flow rate should not adversely affect the reactor, although it might cause overheating in certain heat recovery equipment. Heat recovery equipment will use the incoming stream to cool the exit stream so a sudden decrease in flow might result in transient overheating of the heat exchangers. This is best controlled by careful temperature monitoring and flow control. Large variations in the organic content of the influent may affect the reactor's temperature since the organic materials provide a significant fraction of the heat needed by the system. This factor is best controlled by maintaining a large (on the order of a one-to two-hour supply) holding tank where influent is mixed prior to injection into the SCWO system.

4.6.4 Operations Monitoring

SCWO systems require a high degree of process control for proper operation. The systems should be controlled and monitored by an advanced process control system which incorporates an automatic waste-feed shutoff and orderly shutdown in case of process failure or excursion of key control parameters. In addition to the automatic controls, proper operator training and experience is essential. Items that must be monitored during operation are typical of a high-pressure industrial process and include the following:

- liquid flow rates;
- inlet and outlet absolute pressure for the reactor and all pieces of high-pressure equipment. Interlocks must be installed to stop the influent flow in case of exceedences;
- pressure drop across the reactor and all other pieces of equipment to identify possible plugging. The pressure drop must be monitored independent of the inlet and outlet pressures since the very high overall pressure would tend to mask the gradual increase in pressure drop that might be caused by blockage. Interlocks must be installed to stop the flow in case pressure drops exceed design levels;
- safety shutoffs;
- tank or other storage systems for holding the influent during a shut-down; and
- reactor inlet, outlet, and intermediate temperatures.

4.6.5 Quality Assurance/Quality Control (QA/QC)

No unusual QA/QC procedures are required for SCWO; however, periodic nondestructive testing of the high-pressure components is necessary to ensure safety and operability. All ANSI and ASME codes for high pressure/temperature equipment design, testing, operation and maintenance must be scrupulously adhered to and an appropriate record of compliance and testing must be maintained.

4.7 Case Histories

A number of groups were identified who are currently actively conducting research or marketing commercial SCWO systems. It is impractical to cite all activities in this field herein; however, the following three groups were identified with ongoing commercial activities and technology.

4.7.1 Commerical Activities

4.7.1.1 Eco Waste Technologies Inc.

Eco Waste Technologies (EWT) currently operates a 151.41 L/hr (40 gal/hr) pilot plant at the University of Texas at Austin that has been used for treatment of a wide variety of wastes in numerous tests. A commercial SCWO plant went on-line in the fall, 1995 at Huntsman's Corporation, Austin, Texas plant. The facility is based on the technology developed by EWT and has a capacity to process 27,254.48 L/day (7,200 gal/day) of wastewater. The combined waste feeds to the system consist of process wash water and other chemical wastes containing methanol, polyols, amines, ammonia, and oxygenated organic compounds. The combined stream is about 10% (by weight) organic compounds with a total organic carbon (TOC) content of about 50,000 mg/L (Weismantel 1996). The plant's effluent is claimed to be of consistently high quality. Tests of its performance are planned in Sweden in the near future.

4.7.1.2 General Atomics

General Atomics (GA)[also MODAR, Inc. which was recently acquired by GA, and Organo, the licensee in Japan] currently operates three pilot-scale test systems. The work is being supported by the Environics Directorate of Armstrong Laboratory at Tyndall Air Force Base in Florida. The first pilot plant has a capacity of 5.68 L/min (1.5 gal/min) and is rated for operation at a maximum temperature and pressure of 650°C (1,202°F) and 306 atm (4,500 psig), respectively. It is located at the GA facility in San Diego and was developed for the Defense Advanced Research Project Agency (DARPA) for the treatment of chemical agents, propellants, and other hazardous wastes. This pilot plant has been used most recently for development tests leading to a shipboard SCWO system for destruction of Navy excess hazardous materials. The second pilot plant, located at a site near Brigham City, Utah shown in Figure 4.8 was developed to treat up to 20.9% (by weight) of hydrolyzed solid rocket propellant in aqueous solution. Operating ranges that have been tested are from 425 to 610°C (797 to 1,130°F) and flow rates of 1.14 to 1.89 L/min (0.3 to 0.5 gal/min). The third pilot plant, also located at the GA facility, was acquired during the recent acquisition of MODAR. This pilot plant had been used extensively by MODAR to treat a wide variety of chemical plant wastes in demonstrations conducted for industrial plants including the following.

Figure 4.8
Air Force SCWO Pilot Plant

The reactor is in the "slatted box" in the background, the liquid oxygen tank in the center of the photograph, and the heat exchange systems is to the left outside the area of the photograph.

Reproduced courtesy of General Atomics

- Permitted and developed a process design for a 18,926.72 L/day (5,000 gal/day) demonstration unit to handle a variety of wastes at an operating RCRA-permitted TSD facility in Texas.
- Fabricated a skid-mounted pilot unit for treatment of PCBs, oils, solvents, and sludges at Niagara Falls, NY (G.T. Hong "Hydrothermal Oxidation: Pilot Scale Operating Experiences" Paper No. IWC-95-51, presented at the 56th Annual International Water Conference, October 30-November 1, 1995, Pittsburgh, PA). This unit was also used for treating solvents, biological wastes, and ammonia at a Smith Kline and French manufacturing facility in Pennsylvania (Johnson, J.B., Hannah, R.E., Cunningham, V.L., Daggy, B.P., Sturm, F.J., and Kelly, R.M., "Destruction of Pharmaceutical and Biopharmaceutical Wastes by the MODAR SCWO Process." *Biotechnology*, Volume 6, pp 1423-1427, December, 1988).
- Tested more than 400 pure and complex mixtures in bench and pilot-scale SCWO systems.
- Evaluated more than 100 metals, ceramics, and coatings during more than 5,000 hours of operation under supercritical conditions over the past 14 years.
- Developed and piloted a patented concept for the control and handling of sticky solids generated during the processing of halogenated feedstocks.
- Adapted, implemented, and debugged commercial off-the-shelf hardware and software for computer control of an operating pilot plant.
- Permitted SCWO processes for operation in New York and in Massachusetts.
- Under subcontract to Stone & Webster, GA/MODAR completed a program for the U.S. Department of Energy (DOE) to demonstrate the MODAR-designed pilot plant's ability to effectively process mixed (radioactive and hazardous) wastes representative of those generated in the DOE Weapons Complex. This program, which started in September 1994, included modifications to the existing 1,892.67 L/day (500 gal/day) pilot plant to improve the process performance and facilitate the processing of Trimsol oil contaminated with a variety of heavy metals (Bettinger, Ferland, and Killilea 1994).

4.7.1.3 Sandia National Laboratories

Sandia National Laboratories (Livermore, CA) in cooperation with Foster Wheeler Development Corporation and Aerojet GenCorp currently operates four supercritical water oxidation reactors, two flow, one batch reactor, and one reaction cell for studying hydrothermal flames at supercritical conditions. Further descriptions are now provided for the flow reactors: the Engineering Evaluation Reactor (EER) and the Supercritical Fluids Reactor (SFR). The EER is a second generation reactor developed to evaluate the engineering aspects of SCWO technology. Its modular design facilitates different test configurations. The EER has a maximum operating temperature of 650°C (1,202°F) at an operating pressure of 345 bar (5,000 psi). The total flow capacity is 30 cc/sec (28.5 gal/hr). The system can provide either air or hydrogen peroxide as the oxidizer.

Sandia is working with Foster Wheeler Development Corporation and Aerojet GenCorp to develop the transpiring-wall platelet SCWO reactor. A quarter-scale version of a pilot size unit is operating at Sandia's EER. The test reactor was designed and fabricated by Aerojet. The reactor is protected by transpiration water, which is fed in at a maximum temperature of 450°C (842°F) and the reactor operating pressure. The waste feed stream enters the reactor at subcritical temperature through an injector. The injector mixes the waste, oxidizer, auxiliary fuel, and heating water streams and utilizes the rapid heat release of the auxiliary fuel to initiate oxidation of the waste stream. The mixture is brought to supercritical, reacting conditions. Reaction temperatures of 650°C (1,202°F) are typical.

The SFR is a tubular SCWO reactor, rated for operation at a maximum temperature of 650°C (1,202°F) and pressure of 435 bar (6,300 psi). Mass flow rates of 0.17 to 2.0 g/sec are achievable with residence times of 2.2 to 250 sec. The system is configured to provide isothemal conditions for the controlled assessment of chemical kinetics. In addition, the SFR has a moveable optical cell that can be placed at various positions along the reactor's length. This allows in situ Raman spectroscopic evaluation of chemical constituents within the reactor. Other analytical methods used for effluent analysis include Total Organic Carbon analysis, gas chromatography, ion probes, and turbidity measurements.

Research into SCWO has been conducted by numerous other groups. Examples include Idaho National Engineering Laboratory (INEL) and Los Alamos National Laboratory, both of the Department of Energy.

The following case studies are representative of pilot-scale SCWO activities. They were selected for discussion herein because the tests were conducted under the auspices of an independent third party or because the detailed results have been subjected to peer review. There are numerous other examples of successful application of the technology. While the examples given herein are of pilot-scale programs, the results could be readily scaled up to operation in the 37.85 L/min (10 gal/min) range, a size which is considered to be commercial scale for a system which is designed to treat wastes with very high organics content.

4.7.2 Laboratory-Scale Study, Chemical Agent Treatment

This section presents a set of tests conducted by General Atomics at approximately 50 mL/min on chemical warfare agents (GB, VX, and mustard) and a second set of tests conducted at approximately 1.51 L/min (0.4 gal/min) treating effluent from base hydrolysis of rocket propellant.

4.7.2.1 GB Agent Treatment

The tests were conducted at the Illinois Institute of Technology Research Institute (IITRI), a facility especially designed and certified to safely handle chemical agents, beginning on May 10, 1993. All tests were performed at GB concentrations of approximately 1% (by weight) with 100% excess oxygen. Table 4.7 presents the GB-test matrix.

Table 4.7
GB Agent Bench-Scale Test Matrix

Test No.	Pressure (psig)	Temperature (°C)	Total Flow Rate (mL/min)	Residence Time (sec)	Test Duration (min)
1	4000	550	50.5	16	42
2	4000	450	39.4	29	15
3	4000	550	50.5	16	54
4	4000	450	43.4	26	71
5	4000	500	46.4	20	58

Gas and liquid samples were collected and analyzed throughout the test series. No agent was detected in any liquid samples, signifying a destruction in excess of 99.99999%. Higher destructions have been achieved, but 99.99999% is the maximum that can be measured given an influent agent concentration of 10,000 ppm (1% by weight) and the 1 ppb detection limit for GB in liquid samples. Additionally, no agent above the allowable exposure limit (AEL) was found in the gaseous effluent samples analyzed on-line by a Minicams® analyzer.

After the gaseous and liquid effluent samples were confirmed to be agent-free, they were shipped to the Institute of Gas Technology (IGT) and the University of Texas J.J. Picklo Research Campus (UTPRC), respectively, for further analysis. Gas samples were found to contain oxygen, nitrogen, argon, carbon dioxide, and trace amounts of methane. Liquid samples showed essentially quantitative conversion of the GB agent to complete oxidation products, i.e., hydrofluoric and phosphoric acids. Small amounts (<400 ppm) of methyl phosphonic acid (MPA) and acetone were detected in the samples from the 450°C (842°F) reactor temperature with significantly less detected in the samples from the 500 and 550°C (932°F and 1,022°F) reactor conditions. Table 4.8 shows the analytical results for the GB test series.

4.7.2.2 VX Agent Treatment

VX agent bench-scale testing commenced on June 29, 1993. Six separate tests were performed, investigating temperatures of 450 to 550°C (842°F to 1,022°F)(see Table 4.9). As with the GB test series, all tests were performed at agent concentrations of approximately 1% (by weight) with 100% excess oxygen. Typically, four liquid and two gas samples were taken for each test. Additionally, 4-5 Minicams® analyses of the gaseous effluent were performed during the course of testing. No agent was detected in any liquid samples, signifying destruction in excess of 99.99999%. Higher destructions may have been achieved, but 99.99999% is the maximum that can be measured given an influent agent concentration of 10,000 ppm (1% by weight) and the 1 ppb detection limit for VX in liquid samples. No agent above the allowable exposure limit for VX was detected during on-line Minicams® analysis of the SCWO gaseous effluent.

Gas and liquid samples were sent to IGT and UTPRC, respectively, for further analysis following verification by IITRI personnel of the absence of detectable agent. The VX agent was essentially quantitatively converted during testing to sulfuric and phosphoric acid. Small quantities of transformation products such as acetic acid and acetone were observed. Gas analyses showed the presence of N_2O and very low concentrations of NO_x and SO_x. The results of gas and liquid analyses are summarized in Table 4.10.

Table 4.8
Analytical Results for GB Agent Bench-Scale Tests[a,b]

	Component	Unit of Measure	Reaction Temperature 450°C	500°C	550°C
Liquid	Fluoride	ppm	1431	1719	1216
Analysis	Phosphate	ppm	6524	7785	5683
	MPA	ppm	122	83	48
	Acetone	ppm	385	106	<2
Gaseous	CO	vppm	BDL	BDL	BDL
Analysis	H_2	vppm	BDL	BDL	BDL
	Methane	vppm	BDL	300	100

BDL below detection limit
vppm parts per million by volume

[a]DRE of >99.99999% achieved for all test samples.
[b]Effluent data are averaged for multiple runs performed at the same temperature.

Table 4.9
VX Agent Bench-Scale Test Matrix

Test No.	Pressure (psig)	Temperature (°C)	Total Flow Rate (mL/min)	Residence Time (sec)	Test Duration (min)
1	4000	500	50.5	18	68
2	4000	450	39.4	29	45
3	4000	500	50.5	18	93
4	4000	550	49.5	16	44
5	4000	450	39.4	29	56
6	4000	550	50.5	16	48

Table 4.10
Analytical Results for VX Agent Bench-Scale Tests[a,b]

	Component	Unit of Measure	Reaction Temperature		
			450°C	500°C	550°C
Liquid Analysis	Acetate	ppm	83	446	160
	Phosphate	ppm	3507	3069	3862
	Sulfate	ppm	3474	3013	3812
	Acetone	ppm	22	3	7
	Ammonia	ppm	0	8	0
Gaseous Analysis	CO	vppm	BDL	BDL	BDL
	H_2	vppm	BDL	BDL	BDL
	N_2O	vppm	3000	3500	3700
	NO_x	vppm	<2	<3	4
	SO_x	vppm	4	<4	2

BDL below detection limit
vppm parts per million by volume

[a]DRE of >99.99999% achieved for all test samples.
[b]Effluent data are averaged for multiple runs performed at the same temperature.

4.7.2.3 Mustard Agent Treatment

Unlike GB and VX, mustard agent has a low solubility in water. GB and VX agents were mixed with the water/oxidizer solution in line just upstream of the reactor. They readily went into solution to yield a uniform, miscible solution. Because of its low solubility, mustard agent would yield a two-phase mixture upon feed to the SCWO reactor if fed in the same manner as the VX and GB agents. The two-phase mixture could present operational and performance difficulties in the compact test rig. Therefore, steps were taken to solubilize the mustard agent by hydrolyzing in hot water, thereby producing a uniform feed for SCWO testing.

A small test sample (2 g) was mixed with water in a 10% (by weight) solution and heated. The mixture initially showed distinct water and mustard agent phases, but after heating to 80-90°C (176-194°F) for 7 minutes, it was converted it to a single, clear, uniform phase. A hydrolysis apparatus was assembled to allow controlled hydrolysis of larger samples. Hydrolysis testing showed that mustard agent hydrolysis in water can be completed within approximately 5 minutes at 80 to 100°C (176 to 212°F), if suitably agitated. Hydrolysis at 60°C (140°F) requires approximately four times longer. Following hydrolysis, the solution was cooled and stored for later SCWO use.

Mustard agent testing began on April 5, 1994. Five separate tests were performed, investigating temperatures of 450 to 550°C (842 to 1,022°F)(see Table 4.11). As before, all tests were performed at agent concentrations of approximately 1% (by weight) with 100% excess oxygen. Typically, four liquid and two gas samples were taken for each test. Additionally, multiple Minicams® analyses of the gaseous effluent were performed during the course of testing.

Table 4.11
Mustard Agent Bench-Scale Test Matrix

Test No.	Pressure (psig)	Temperature (°C)	Total Flow Rate (mL/min)	Residence Time (sec)	Test Duration (min)
1	4000	450	33.3	34	55
2	4000	450	31.3	37	55
3	4000	550	aborted	aborted	aborted
4	4000	500-525	30	34	92
5	4000	500	32.5	28	47
6	4000	500	31.5	29	47

No agent was detected in any liquid samples, signifying destruction in excess of 99.9999%. Higher destructions may have been achieved, but 99.9999% is the maximum that can be measured given an influent agent concentration of 10,000 ppm (1% by weight) and the 10 ppb detection limit for mustard agent in liquid samples. Also, except for Runs 3 & 4, where a system upset resulted in agent contamination of the effluent collection lines, no readings above the mustard agent AEL of 0.003 mg/m^3 were detected. During Run 3, an equipment malfunction necessitated test termination. Residual feed material was flushed at reduced pressure and temperature through the reactor into the effluent collection lines, thus contaminating them. This was not discovered until after the start of Run 4.

Gas and liquid samples were sent to specialized laboratories for further analysis. The results are listed in Table 4.12. The data for runs performed at the same temperature have been combined and averaged. Meaningful chloride measurements could not be made because sodium chloride has previously been added to the wastes in order to stabilize them for storage and shipping. Low levels (≤600 ppm) of the intermediate transformation products acetic and formic acids were observed at 450°C (842°F), with less observed at 500°C (932°F), and none observed at 525°C (977°F). Relatively high concentrations of CO were observed at lower temperatures, decreasing to less than 2,000 ppm at 525°C (977°F). Higher operating temperatures will even further reduce observed CO levels.

Table 4.12
Analytical Results for Mustard Agent Bench-Scale Tests[a,b,c]

	Component	Unit of Measure	Reaction Temperature 450°C	500°C	525°C
Liquid	Acetate	ppm	597	47	<1
Analysis	Formate	ppm	211	16	<1
	Sulfate	ppm	6008	5336	3105
Gaseous	CO	mole%	14.5	1.4	0.17
Analysis	H_2	vppm	3000	BDL	BDL
	C_2H_4	vppm	700	BDL	BDL
	SO_x	vppm	2133	1098	2600

BDL below detection limit
vppm parts per million by volume

[a]DRE of >99.99999% achieved for all test samples.
[b]Effluent data are averaged for multiple runs performed at the same temperature.
[c]Several unknown, nonagent peaks of ~50-100 ppm each were detected.

SO_x levels of 1000 to 2600 ppm were observed during testing. A major factor contributing to these relatively high levels is thought to be poor mixing/mass transport limitations caused by the low flow rates and short lengths of the test system. Even so, the sulfur present in the observed SO_x represented only about 2% of the available sulfur, the remaining having been fully converted to sulfate. Improved mixing in a redesigned pilot-plant reactor should result in lower SO_x levels.

The DARPA HTO (High Temperature Oxidation) pilot plant is designed to provide a transportable pilot-scale demonstration unit for extremely hazardous wastes such as chemical warfare agents, as well as solid propellants and other U.S. Department of Defense (DoD) hazardous wastes. The system is designed to provide a very flexible test bed. The maximum operating pressure and temperature are 310 bar (4,500 psi) and 650°C (1,202°F). It has a nominal flow rate of 3.79 L/min (1 gal/min) with a typical feed concentration of 5% (by weight) organics and up to 12% (by weight) inert solids.

All of the high-pressure equipment, other than the oxygen supply system and the high-pressure water pump, are contained on the reactor skid. There are numerous flanges in the lined system to allow installation of instruments and special test equipment. The system can be operated with or without heat recovery and preheaters. The reactor skid is enclosed with polycarbonate shielding to provide personnel protection from all high pressure components and to contain the effluent in the event of a system rupture prior to discharge to the facility ventilation system.

In the fall of 1996, this pilot plant was used to test the ability of SCWO to treat hydrolized VX for the U.S. Army. The tests were considered successful with destruction of organic compounds achieved to no measurable VX-thiol, the most prevalent VX compound present in the hydrolysate. The detection limit for these measurements is equivalent to 99.9999% destruction. Based on the results of these tests in early 1997, the Army selected hydrolysis followed by SCWO treatment of the hydrolysate as the method to be used to treat VX, GB and mustard stockpiles around the U.S. The full-scale system is currently being designed.

4.7.3 Pilot-Scale Studies

4.7.3.1 Hydrolyzed Rocket Propellant Treatment

For the Air Force, the prototype hydrolysis and HTO systems built by General Atomics were installed at a Thiokol site near Brigham City, Utah where demonstration tests were recently completed. The Air Force prototype HTO system installed at Thiokol has a maximum operating pressure of 310 bar (4,500 psi), a maximum temperature of 650°C (1,202°F) and a nominal flow rate of 1.59 L/min (0.42 gal/min). All of the high pressure equipment, other than those associated with the oxygen supply system, are contained on the reactor skid. Like the ARPA system, there are numerous flanges in the lined piping to allow installation of instruments and special test equipment. The reactor skid is enclosed with polycarbonate shielding to provide personnel protection from all high pressure components. A RCRA RD&D permit was obtained for the Air Force system at Thiokol in a period of about six months.

Two test runs were conducted on effluent from the base hydrolysis of rocket propellant over a wide range of operating conditions. The initial test processed ~25 pounds of double base hydrolyzed propellant as a 1% (by weight) hydrolyzed solution. The result of a continuous run exceeding 24 hours was total organic carbon (TOC) destruction to below the detection limit.

A second 695 lb batch of hydrolyzed double base propellant at a concentration up to 21% was processed during a continuous 34-hour run and demonstrated reliable system performance over the range of temperatures from 450 to 580°C (842 to 1,076°F) and flow rates of 1.14 to 1.67 L/min (0.3 to 0.44 gal/min).

Effluent sample analysis confirmed pretest predictions of hydrolyzed effluent treatment with TOC and NO_x levels below the detection limit (1 and 5 mg/L, respectively) for operating temperatures in the 570 to 580°C (1,058 to 1,076°F) range at a pressure of 276 bar (4,000 psi).

Propellant throughput rates of up to 800 lb/day (24-hr/day equivalent) were demonstrated over several hour run times at several operating points. Instantaneous propellant throughput rates, extrapolated to 24-hr/day operation, of 1,100 lb/day were achieved. According to General Atomics, the operator, with fuel additives, these rates could potentially be significantly higher. No evidence of corrosion was found in the titanium-lined sections of the preheater, reactor, and cool-down heat exchangers. The high solids content feed was processed with stable pressure control and with no evidence of plugging or erosion. The system operated within the requirements specified by the RCRA RD&D permit obtained for the SRMD Prototype Facility for processing of hazardous wastes.

4.7.3.2 Paper Mill Wastes Treatment

The University of Texas at Austin currently owns and operates an approximately 227.12 L/hr (60 gal/hr) capacity, continuous-flow SCWO system (Blaney et al. 1995). Kimberly-Clark Corporation has performed numerous experiments at this facility and has shown that paper mill sludges can be converted (oxidized) to clean water, clean calcium carbonate ash, and clean gas (carbon dioxide and residual oxygen). Experiments were performed on de-inking sludge from a paper recycling operation. Treatment results for virgin sludges are reported elsewhere (Hossain 1991).

The paper mill wastes contained small traces of polychlorinated dibenzodioxins and dibenzofurans (PCDD/PCDF). The SCWO system's ability to destroy these compounds were of particular concern during these tests. Two tests were conducted at the following SCWO reactor conditions:

- pressure — 245 atm (3,600 psi)
- flowrate — 75.71 L/hr (20 gal/hr)
- reaction temperature — 450 and 500°C (842 and 932°F)

The overall organic destruction was quantified in terms of total organic carbon (TOC). The thoroughness of destruction of trace chlorinated organics was quantified by measuring the trace quantities of PCDDs, PCDFs, and PCBs before and after SCWO, using gas chromatography and high resolution mass spectrometry following standard methods specified by US EPA (SW-846).

The major components in the pilot plant consisted of a high pressure diaphragm pump, a number of double pipe heat exchangers, an electric heater, a coiled tube reactor, an air-driven oxygen booster, and a hydrocyclone. The heat exchanger between the feed and reactor effluent allowed the recovery of some thermal energy, reducing the electrical power input required to maintain the process. The oxygen flowrate, electrical power input to the heater, and process pressure and temperature were controlled and monitored by a computer. The feed flowrate was controlled by manually adjusting the stroke length dial at the pump head and was monitored by the computer. The reactor volume was 13.08 L (49.51 gal).

The feed was aqueous de-inking sludge collected at a paper recycling mill before biological treatment. It contained 5 to 7% solids, the solids portion being 30% inorganic (clays, fillers) and 70% organic (cellulosic fiber fines, residual pulping and de-inking chemicals, and trace chlorinated organics). The sludge pH was about 5.7, and total chlorides about 1.1 mg/L. The sludge was diluted to 2.5% solids, and the diluted mixture, termed the "feed sludge," was subjected to SCWO treatment. In order to increase the feed sludge's heating value for the higher-temperature test, Experiment B, a small amount of methanol was added to the feed holding tank and blended evenly throughout the feed sludge. Table 4.13 presents the feed sludge values for total organic carbon (TOC), PCDDs, PCDFs, and PCBs on a dry basis, not including the added methanol.

Table 4.13

Composition of Feed Sludge and Product Ash[1]

	Feed Sludge[2]	Expt. A Product Ash[3]	Expt. B Product Ash[3]
TOC	440,000 ppm	3400 ppm	<100 ppm
2,3,7,8-TCDD	3.99 ppt	<0.325 ppt* (DE>99%)	<0.23 ppt* (DE>99%)
Total TCDD	6.41 ppt	18 ppt	1.5 ppt
Total PeCDD	<2.46 ppt*	<2.5 ppt*	<0.81 ppt*
Total HxCDD	15.36 ppt	<1.23 ppt*	<0.445 ppt*
Total HpCDD	33.81 ppt	5.25 ppt	<0.64 ppt*
OCDD	470.70 ppt	<7 ppt*	1.6 ppt
Total PCDD	530 ppt	23.3 ppt (DE>96%)	3.1 ppt (DE>96%)
2,3,7,8-TCDF	35.09 ppt	2.85 ppt (DE>99%)	<0.505 ppt* (DE>99%)
Total TCDF	72.42 ppt	63.5 ppt	7 ppt
Total PeCDF	9.37 ppt	<1.05 ppt*	<0.275 ppt*
Total HxCDF	<2.97 ppt*	<0.8 ppt*	<0.34 ppt*
Total HpCDF	12.06 ppt	<1.19 ppt*	<0.22 ppt*
OCDF	15.19 ppt	<1.5 ppt*	<0.53 ppt*
Total PCDF	144 ppt	66.4 ppt (DE>96%)	7 ppt (DE>96%)

Table 4.13 cont.
Composition of Feed Sludge and Product Ash[1]

	Feed Sludge[2]	Expt. A Product Ash[3]	Expt. B Product Ash[3]
mono-chlor-biphenyl	<0.32 ppb*	46.5 ppb	<0.2 ppb*
Di-Chl-Bp	8.65 ppb	20.5 ppb	<0.2 ppb*
Tri-Chl-Bp	<1.51 ppb*	<0.945 ppb*	<0.2 ppb*
Tetra-Chl-Bp	<0.65 ppb*	<0.4 ppb*	<0.4 ppb*
Penta-Chl-Bp	<0.69 ppb*	<0.4 ppb*	<0.4 ppb*
Hexa-Chl-Bp	<0.65 ppb*	<0.4 ppb*	<0.4 ppb*
Hepta-Chl-Bp	<1.29 ppb*	<0.8 ppb*	<0.8 ppb*
Octa-Chl-Bp	<1.29 ppb*	<0.8 ppb*	<0.8 ppb*
Nona-Chl-Bp	<3.23 ppb*	<2 ppb*	<2 ppb*
Deca-Chl-Bp	<3.23 ppb*	<2 ppb*	<2 ppb*
Total PCBs	8.65 ppb	67 ppb (PCBs formed)	<DL*(DE>90%)

TCDD tetrachlorodibenzo-para-dioxin
PeCDD pentachlorodibenzo-para-dioxin
HxCDD hexachlorodibenzo-para-dioxin
HpCDD heptachlorodibenzo-para-dioxin
PCDD polychlorinated dibenzo-para-dioxin
PCDF polychlorinated dibenzofuran, etc.
ppb parts per billion (μg/L)
ppt parts per trillion

[1]All data reported on dry basis
[2]Average of 4 analyses
[3]Average of 2 analyses
*Below detection limit

Reprinted with permission from Hutchenson and Foster, *Innovations in Supercritical Fluids*, Series 608, p 448. Copyright 1995 American Chemical Society.

The feed sludge was pressurized to approximately 245 atm (3,600 psig) using a high-pressure diaphragm pump, and then heated with electrical heaters to reach the desired reaction temperature. While the heating value of the sludge was sufficient to maintain the reactor's temperature, poor insulation of the pilot-scale test reactor required additional heat which was provided by electrical heaters and a number of double-pipe heat exchangers recovering energy from the processed reactor effluent.

Figure 4.9
SCWO Reactor Temperature Profile

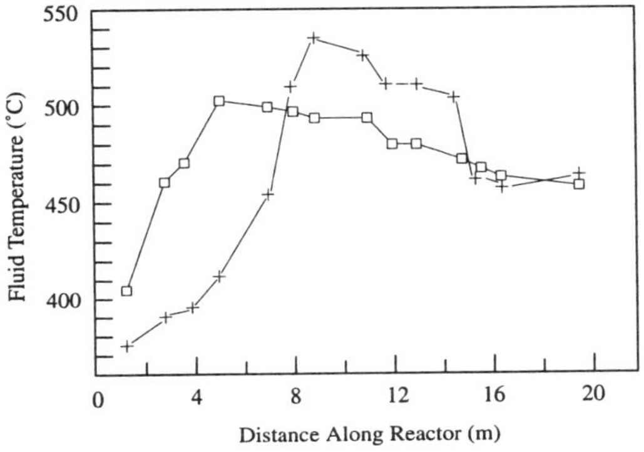

□ Experiment A
+ Experiment B

Reprinted with permission from Hutchenson and Foster, Innovations in Supercritical Fluids, Series 608, p 452. Copyright 1995 American Chemical Society.

The influent flowrate was 75.71 L/hr (20 gal/hr). Reactor temperature profiles for the experiments are shown in Figure 4.9. The residence time at reactor conditions was about 50 seconds. Two sets of tests were conducted at the following conditions:

Experiment A. Average reactor temperature was maintained at approximately 450°C (842°F), all other variables as described above.

Experiment B. Average reactor temperature was maintained at approximately 500°C (932°F).

For Experiment B, 2% (by weight) methanol was mixed with the feed in the influent holding tank. The methanol acted as an auxiliary fuel and its added heat of combustion brought the reactor temperature to 500°C (932°F). Since methanol is very easily oxidized to carbon dioxide and water at SCWO conditions, it is assumed that adding this small amount of methanol would only minimally affect the product mix.

Feed, product water, and product ash from both Experiment A and Experiment B were analyzed within two weeks of the trials. Analyses were for all congeners of PCDDs, PCDFs, and PCBs, Total Organic Carbon (TOC), percent solids, and metals. The analyses were conducted at an independent laboratory which utilized gas chromatography and high-resolution mass spectrometry for trace chlorinated organics analyses following standard US EPA procedures. In addition, numerous analyses of the aqueous phase for suspended solids, chemical oxygen demand, acetic acid, ammonia, chlorides, and pH were performed. Tests on the solid (ash) phase included volatile solids and the Toxicity Characterization Leaching Procedure (TCLP).

Data for the product streams are shown in Table 4.13 adjacent to the feed data. Values of PCDDs, PCDFs, and PCBs in the product water were less than one percent of that in the product solids (ash) so these data were not included. For both experiments the majority of the PCDDs and PCDFs were destroyed.

In Experiment B, at 500°C (932°F), the oxidation appeared to approach completion, as the TOC in the solids was under the detection limit of 100 mg/L, and virtually all of the PCDDs and PCDFs were destroyed. Destruction Efficiencies (DEs) of 2,3,7,8- TCDD and 2,3,7,8-TCDF (widely accepted as the most toxic congeners) were over 99%, and DEs of total PCDDs and PCDFs were over 96%. Over 90% of the PCBs were destroyed.

In Experiment A, the destruction efficiencies for PCDDs and PCDFs were roughly the same as for Experiment B (>99%). However, at the lower temperature of 450°C (842°F) in Experiment A, the data indicate that PCBs form in the parts per billion level. In addition, TOC destruction was unacceptably low. SCWO treatment reduced the TOC in the product ash to 3,400 mg/L from 440,000 mg/L in the feed solids. Possible PCB formation was not observed at the higher SCWO temperatures and TOC destruction was also greater. The formation of PCB was clearly significant. No other literature was found showing such results and the results were not replicated at the higher temperatures. Site-specific testing is needed to confirm and expand upon these findings. Until these results are confirmed or refuted it appears prudent in those applications where chlorinated organic compounds are treated to operate at temperatures above 500°C (932°F) and to test the effluent for PCB.

At first glance it may appear that some TCDDs were formed in experiment A; this is not the case. A complete mass balance which takes into account the fact that the 18 ppt total TCDD in the product ash is based on inorganic only (dry basis), whereas the 6.41 ppt total TCDD in the feed solids is based on a mixture of inorganic plus organic, also on a dry basis.

The solid residues (ash) derived from the SCWO tests were characterized by the Toxicity Characteristic Leaching Procedure. Concentrations of the metals in the leachate from the ash were lower than the regulatory levels set by the US EPA. Most heavy metals, including As, Cd, Cr, Hg, Ni, Pb, Se, Tl, and V, were nonleachable (below the detection limit of 0.0005 mg/g solid).

Supercritical water oxidation at temperatures of 500°C (932°F), pressures of 245 atm (3,600 psi), and a residence time of about 50 sec, was shown to be effective in destroying over 99% of the most toxic dioxin-type congeners, 2,3,7,8-TCDD and 2,3,7,8-TCDF, over 96% of the total PCDD/PCDFs, and over 90% of the PCBs. However, at a lower temperature of 450°C (842°F)(other conditions remaining constant), destruction of chlorinated organics was not as thorough, and PCBs may have actually formed and survived for a short time.

4.8 Conclusion

SCWO appears to be a technology that is reaching commercial scale. Destruction efficiencies for organic materials equivalent to those achieved by incinerators have been demonstrated and, sufficient knowledge appears to have been accumulated to scale (at least some SCWO system designs) to the 10 gallon per minute range which the authors generally consider to be commercial scale. The cost of treatment is still projected to be somewhat higher than for incineration so that the initial applications will, most likely, be for the treatment of specialized wastes, such as rocket propellant or chemical warfare agents which (because of technical or social objections) cannot be incinerated. It appears likely that additional experience will reduce the cost and this technology should be considered for those applications where the waste includes aqueous streams contaminated with high concentrations of organic compounds.

5

EX-SITU HIGH VOLTAGE ELECTRON BEAM TREATMENT

5.1 Introduction

The electron beam (E-beam) technology is a means of treating wastewaters and groundwaters contaminated with organic compounds. The technology was developed by High Voltage Environmental Applications, Inc. (HVEA) of Miami, Florida, in conjunction with Florida International University and the University of Miami. A full-scale system utilizing the E-beam technology was installed at the Miami-Dade Water and Sewer Authority Wastewater Treatment Plant in Key Biscayne, Florida in approximately 1983 as a means of sterilizing the sewage sludge from the wastewater treatment plant (Kurucz, Waite, and Cooper 1995). No longer used for this purpose, it serves as an experimental unit for research and treatability studies. This fixed system has a capacity of 460 L/min (120 gal/min) and uses a 1.5 MeV, 50 mA (75 kW) electron accelerator as the radiation source for treatment.

HVEA operates a second, mobile pilot-treatment system (model M25W-48S) with a capacity of 19 to 190 L/min (5 to 50 gal/min) which has been used for a number of demonstrations in the U.S. and Europe. This system uses an electron beam with a maximum power output of 25 kW (US EPA 1995d). The majority of the information presented is based on pilot-scale demonstrations using this mobile system.

A bench-scale batch treatment system with a capacity of approximately two gallons of wastewater is also available for treatability studies. The bench-scale system uses gamma as a source of electrons for the beam produced by ^{60}Co jacketed in stainless steel. The ^{60}Co produces, on decay, one beta particle and two gamma rays. The stainless steel is constructed to stop the beta particles, allowing only the highly-penetrating gamma rays to escape into the surrounding medium and irradiating the material to be treated (US EPA 1995d; Kalen 1992).

5.2 Process Description

High voltage electron beam treatment (HVEBT) treats aqueous streams contaminated with organic constituents or with pathogens by irradiating the stream with a high-energy electron beam. Treatment is conducted at normal temperature and pressure. Figure 5.1 is an elevation drawing of the 75 kW facility in Key Biscayne, Florida. Physically, the system is simple. There are no moving parts, except for standard pumps. The aqueous stream is passed over a weir (influent spreader in Figure 5.1) which converts it into a cascading flat curtain of water. A relatively thin water flow is necessary since the electron beam's treatment efficacy is reduced as the thickness of the water increases. An electron gun similar to the electron gun found in common cathode ray tubes (i.e., television tubes) rapidly scans across the flat curtain of water.

The high-energy radiation sterilizes the stream and destroys organic contaminants by chemical redox mechanisms. Inorganic constituents in the aqueous stream can also be chemically modified, although the data on the behavior of inorganic constituents is not well established. Organic chemical destruction efficiencies exceeding 99% have been achieved. In cases of high organic concentrations in the influent, multiple passes through the electron beams might be required to achieve the desired effluent standards. The irradiation chamber and electron beam source used for treatment must be shielded, but the treatment leaves no residual radioactivity in the aqueous stream.

The interaction between the electron beam and the waste stream being treated raises the temperature of the waste less than five degrees. It appears unlikely, that under normal operating conditions, interaction of the electron beam with ambient air results in the formation of ozone and other trace contaminants (O, CO, and NO_2). However, the small air stream flowing across the titanium window of the electron gun requires some form of treatment prior to discharge, mainly for the catalytic destruction of O_3. Since the volume of air used to cool the titanium window is relatively low, E-Beam treatment systems would require an air pollution abatement system much smaller than those typically utilized by competing technologies.

Figure 5.1
Elevation of the Electron Beam Research Facility, Key Biscayne, Florida

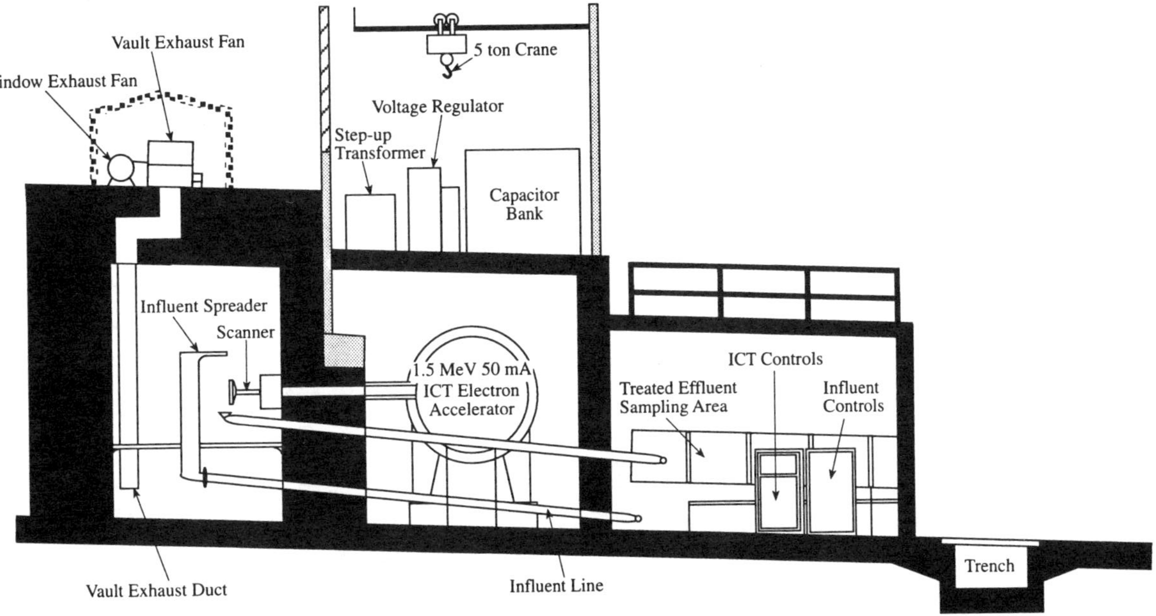

Reproduced courtesy of High Voltage Environmental Applications, Inc.

A potential concern in the use of this treatment method is that the simple flow of the aqueous stream across an open weir can cause a release of volatile contaminants such as trichloroethylene (TCE) and trichloroethane (TCA) into the air within the treatment vessel. This air is normally contained in the treatment chamber; however, during the US EPA-SITE demonstration (US EPA 1995d) of this technology it was found that the cooling air stream flowing past the titanium window isolating the electron beam source was leaking into the treatment chamber and trace amounts of the volatile organic compounds in the wastewater stream were being removed by stripping. The problem was traced to the waste delivery system which allowed waste stream interaction with the cooling air. Interaction between the halogenated VOCs with the ozone and molecular oxygen in the cooling air, and with the electron beam, released the VOCs and, in addition, formed small amounts (low ppm) of HCl and phosgene.

This problem, although environmentally not significant because of the low flow rates and low contaminant concentrations of the leaking stream, has been completely eliminated in newer designs in which, the waste delivery system uses a completely enclosed (proprietary) means to distribute the flow of waste. The new design also isolates the waste from the electron beam source by a second titanium window so that the gas stream that cools the electron source never comes in contact with the waste stream. The new design also recirculates the cooling gas stream, so that the ozone produced by the electrical discharge, is recirculated. According to the vendor, these design modifications eliminate the need for added air pollution control equipment in almost all cases.

Table 5.1 compares the E-Beam technology against a number of alternative technologies for treating water contaminated with low molecular weight organic compounds.

5.3 Scientific Principles

The high energy electron beam passing through the aqueous stream causes atoms in its path to achieve highly-excited electron states and form free radicals. Organic destruction occurs because the electron beam, when it interacts with water, generates both powerful oxidizing radicals such as the hydroxyl radical (•OH) as well as reducing radicals such as the aqueous electron (e^-_{aq}) and the hydrogen radical (H•)(Spinks and Woods 1990). These free radicals react with other constituents in the water. Unlike photochemical reactions where one photon of light initiates one (molecular) reaction, a high energy electron is capable of initiating several thousand reactions as it dissipates its energy (Spinks and Woods 1990).

Table 5.1
Comparison of Technologies for Treating VOCs in Water

Technology	Advantages	Disadvantages
Air Stripping	Effective for high concentrations; mechanically simple; relatively inexpensive	Inefficient for low concentrations; VOCs discharged to air
Steam Stripping	Effective for all concentrations	VOCs discharged to air; high energy consumption*
Air Stripping with Carbon Adsorption of Vapors	Effective for high concentrations	Inefficient for low concentrations; requires disposal or regeneration of spent carbon; relatively expensive
Air Stripping with Carbon Adsorption of Vapors and Spent Carbon Regeneration	Effective for high concentrations; no carbon disposal costs; product can be reclaimed	Inefficient for low concentrations; high energy consumption
Carbon Adsorption	Low air emissions; effective for high concentrations	Inefficient for low concentrations; requires disposal or regeneration of spent carbon; relatively expensive
Biological Treatment	Low air emissions; relatively inexpensive	Inefficient for high concentrations; slow rates of removal; sludge treatment and disposal required
Chemical Oxidation	No air emissions; no secondary waste; VOCs destroyed	Not cost-effective for high contaminant concentrations; may require chemicals such as O_3 and H_2O_2
E-Beam System	No secondary waste; multiple mechanisms for VOC destruction; no chemicals (such as O_3 or H_2O_2) required	High electrical energy consumption; not cost-effective for high contaminant concentrations; relatively expensive**

*Vendor adds that stream stripping does not destroy the contaminant but only removes it from the wastewater into a second stream that then must be disposed.
**This conclusion is given in the US EPA report, but it is questioned by the vendor.

Source: US EPA 1995d

The efficiency of conversion of a high energy electron beam to a chemical process is defined as G, which is the number of chemically active moities (radicals, excited atoms, or other reactive products) formed or lost in a system absorbing 100 electron volts (eV) of energy absorbed. Because water is by far the predominant molecule found in any system likely to be treated, it will form the predominance of reactive moities in a typical system. However, data collected on pure water spiked with selected contaminants will not be generally applicable to actual contaminated waters because other constituents in the influent stream play a significant role in the chemical reaction by scavenging free radicals. Carbonates, iron, other inorganic compounds, and natural and synthetic organic compounds (other than target materials) are examples that may affect destruction efficiency by acting as radical scavengers. As a result, experimental data used for scale-up must be obtained from actual site samples, rather than by spiking a readily available clean water stream.

Because of the aggressive nature of the free radicals and other reactive moities formed by the electron beam, the rates of reactions are rapid. Rate constants (pseudo first-order) for the chemical reactions between the (e^-_{aq}), (H•), and (OH•) free radicals and a variety of organic compounds are in the range of 10^7 to 10^{10} $mol^{-1}sec^{-1}$ (Spinks and Woods 1990).

5.4 Potential Applications

The greater the initial concentration of the organic contaminants in the influent, the greater the dose of high-energy electrons that are required to achieve a given effluent concentration. Therefore, the costs associated with treatment increase substantially as the concentration increases.

The technology is generally applicable to aqueous streams or flowable slurries (3-5% solids). Furthermore, the systems have a high tolerance for suspended solids (Cooper et al. 1992). In general, materials handling constraints and the high electron energies that would be required to penetrate beds of solid materials (e.g., soils) limit the system's usefulness to flowable aqueous matrices.

The greater the organics loading of the influent stream, the greater the quantity of radiation required to achieve a level of treatment and, hence, the greater the energy cost. As a result, cost of treatment increases rapidly as the organic concentration in the influent stream approaches approximately 1%.

High voltage electron beam treatment has been demonstrated to successfully destroy a wide variety of organic compounds dissolved in waters from many different sources. It has successfully treated chlorinated organic compounds (US EPA 1995d). Recently, bromate ions were reduced to bromide ions (Siddiqui et al. 1996) in drinking water. Table 5.2 displays some matrices and compounds that have been treated using this technology (Kurucz et al. 1991b). The table also gives the radiation doses used to achieve this removal. It is noted that recent work has demonstrated higher removal efficiencies for many of these compounds.

While the high energy electron beam process can be used as a stand-alone process, its ability to chemically degrade a large variety of biologically-refractive compounds also make it a candidate for the pretreatment of streams containing these types of compounds.

The pilot- and full-scale application of this process has been used to treat aqueous streams with dissolved organics. Recent work using the full-scale system has shown that electron beam treatment significantly improves the dewatering properties of sewage sludge (Waite et al. 1996). The irradiation enhanced agglomeration by altering the charge on the sludge particles. If these results can be extended to other sludges, the technology may also find application in the dewatering of sludges and sludge-like materials from remediations.

5.5 Treatment Trains

The aqueous stream to be processed usually requires minimum pretreatment. The system has no heat transfer surfaces, tight orifices, windows, or other areas that are subject to fouling. Only a coarse screening to remove debris that may catch on the weir or which might damage equipment as it falls over the weir need to be removed prior to treatment.

Table 5.2
Summary of Percent Removal of Various Organic Compounds by Treatment Application Area

	Percent Removal	Required Dose (Krads)
Drinking Water		
Chloroform	83	650
Bromodichloromethane	>99	80
Dibromochloromethane	>99	80
Bromoform	>99	80
Wastewater/Groundwater Treatment		
Carbon Tetrachloride	>99	50
Trichloroethylene (TCE)	>99	500
Tetrachlorethylene (PCE)	>99	500
Trans-1,2-Dichloroethene	93	800
Cis-1,2-Dichloroethene	98	800
1,1-Dichloroethene	>99	800
1,2-Dichloroethane	60	800
Hexachloroethane	>99	800
1,1,1-Trichloroethane	89	650
1,1,2,2-Tetrachloroethane	88	650
Hexachloro-1.3-Butadiene	98	800
Methylene Chloride	77	800
Groundwater Treatment		
Benzene	>99	650
Toluene	97	650
Chlorobenzene	97	650
Ethylbenzene	92	650
1,2-Dichlorobenzene	88	650
1,3-Dichlorobenzene	86	650
1,4-Dichlorobenzene	84	650
m-Xylene	91	650
o-Xylene	92	650
Dieldrin	>99	800
Total Phenol	88	800

These tests were conducted at 120 gal/min and one pass only. Larger electron beam dosages would result in a greater destruction of the organic compound.

Reprinted from *Advances in Nuclear Science and Technology*, Volume 22, Kurucz et al., p 36, 1991 with permission of Plenum Press.

It should be noted that organic compounds that are ad- or absorbed by grit or other solids in the wastewater will probably not be subjected to the high-energy electron beam, since the inert portion of the solid will absorb some of the beam. As a result, ad- or absorbed contaminants will, most likely, not be destroyed to the same level as will dissolved organics. The vendor claims to have demonstrated acceptable organics destruction on waters containing up to 3% solids.

Waste streams containing concentrations of organic in excess of 1%, or which contain organic materials in suspension, should be pretreated using physical means. Possible pretreatments include gravity separation (oil-water separators), floatation, or flocculation. Because the accelerated electrons actually convert the water into a reagent, this process works best when it only has to destroy dissolved organic compounds.

Posttreatment requirements for the treated effluent from this process depend on the nature of the feedwater and of the discharge requirements for the site. Even if the treated effluent from this process does not meet the site-specific discharge requirements, the electron beam process will, in most cases, "soften" the organic compounds, thus making the stream amenable to biological treatment.

5.6 Design

5.6.1 Design Basis

The key parameter that must be considered in the design of an E-beam treatment system is the energy required to destroy the contaminant in question to the discharge limit. Some guidance on this matter is given in Table 5.1; however, it is necessary to conduct treatability studies in order to establish the necessary electron beam energy and the type and level of posttreatment that might be required. The system's inherent simplicity makes such treatability studies at all scales relatively inexpensive to conduct. The existing pilot-scale unit, for which a large amount of scale-up data is available, is mounted on a trailer and it can be readily moved to the site. Treatability studies using the pilot-scale system have been successfully conducted at a number of sites.

5.6.2 Design and Equipment Selection

Two categories of equipment are needed for this process — water transfer and distribution equipment and irradiation equipment. The water transfer and distribution equipment are simply pumps, a waste delivery system, and other standard water handling equipment. These are readily available or they can be quickly fabricated. The irradiation equipment consists of standard electron beam guns that can also be readily purchased or fabricated. The electron beam equipment used for the full-scale system was built in the early 1970s and is still functional; testimony to its reliability and durability.

5.6.3 Process Modification

The only parts of the process itself that can be modified are the electron source and the water distribution equipment. The electron source can either be an electron gun or a radioactive beta particle source. Cobalt-60, which is radioactive, has been used as an electron (beta-particle) source in the laboratory-scale system, but its use for commercial-scale systems is not economically viable, according to the vendor.

The design of the waste delivery system and the thickness of the sheet of water cascading from the weir is another possible process modification. Reducing the depth of water through which the electrons must penetrate improves the fractional destruction of the organics, although it will not necessarily improve the overall system performance, since the electron utilization rate will decrease. The optimum configuration is best established on the basis of treatability studies.

The removal efficiency for the E-Beam technology increases with multiple passes of the wastewater through the system. A modification of the process that involves changes in the operating procedure is to recycle some or all of the effluent from the process. For total recycle, using the same equipment would require batch treatment of the wastewater. Two holding tanks, one feeding the E-beam system and the second holding the treated effluent, and then reversing their rolls would be needed for such an application. Alternatively, multiple E-beam systems could be assembled in series or in parallel, depending on whether the application required high levels of irradiation (and hence, high destruction) or high flow rates. Another modification for a continuous flow system is to recycle a portion of the effluent.

Using this procedure, removal efficiencies (REs) could be improved without the use of multiple electron beam generators and their inherent energy costs.

5.6.4 Pretreatment Processes

Waste streams containing concentrations of organic in excess of 1%, or which contain organic materials in suspension, should be pretreated using physical means. Possible pretreatments include gravity separation (oil-water separators), floatation, or flocculation. Because the irradiation process actually converts the water into a reagent, this process works best when it only has to destroy dissolved organic compounds.

5.6.5 Posttreatment Processes

Posttreatment requirements for the treated effluent from this process depend on the nature of the feedwater and of the discharge requirements for the site. Even if the treated effluent from this process does not meet the site-specific discharge requirements, the electron beam process will, in most cases, "soften" the organic compounds, thus making the stream amenable to biological treatment.

5.6.6 Process Instrumentation and Control

The processing instrumentation and controls consist of a voltage or current regulator to maintain the electron beam at a constant power, and controls for the electromagnets which are used to cause the electron beam to scan in a controlled pattern across the water flowing through the waste delivery system. This equipment is conceptually identical to the electron beam controls of a simple cathode ray tube, but operate at higher current levels. This control equipment has been commercially available for well over fifty years.

5.6.7 Safety Requirements

The irradiation equipment is a source of ionizing radiation. As a result, appropriate shielding must be incorporated into the design. Fortunately, there is no lingering radiation in the "hot" radiation areas of the system; radiation only occurs when the electron beam is operating. When the beam is off, there is no risk of radiation exposure to personnel entering the irradiation chamber.

Interlocks must be installed on the access doors to the irradiation chamber to prevent entry when the electron beam is energized. Also, an emergency shut-off switch must be installed in the irradiation chamber to allow anyone who is inadvertently trapped within to deactivate the system.

5.6.8 Specification Development

The key requirements that must be incorporated in specifications for an electron beam treatment application depend on whether the bids are for equipment, which is to be installed by others, or for a turnkey electron beam system with performance guarantees.

If a vendor is to provide a turnkey system, then the performance guarantees must be based on quality of the water to be treated. Competing vendors should be supplied with the complete results of all treatability studies conducted.

Specification of individual equipment to be installed is fairly straightforward. However, the special equipment for this process is covered by patents held by the developer and, in general, can only be acquired through this sole-source.

It is important to remember that this process employs the application of high voltage at a large scale and relatively high power under highly humid conditions. Therefore, evaluation of vendor bids should include an evaluation of the vendors' qualifications and experience as well as a comparison of the cost data submitted.

5.6.9 Cost Data

The following analysis presents cost information for using the HVEA E-beam technology to treat groundwater contaminated with VOCs. Cost data were compiled during the Superfund Innovative Technology Evaluation (SITE) demonstration at the Savannah River Site (SRS) and from information obtained from independent vendors and HVEA. Costs are presented in February, 1995, dollars and are considered to be order-of-magnitude estimates with an expected accuracy within 50% above and 30% below the actual costs.

Two models, based on different groundwater characteristics, are presented and compared. In Case 1, the groundwater has an insignificant level of alkalinity (<5 mg/L as $CaCO_3$) and contains VOCs that are easy to destroy using free radical chemistry. In Case 2, the groundwater has moderate-to-high alkalinity (500 mg/L as $CaCO_3$) and contains additional VOCs, a few of which are more difficult to destroy. In Case 1, a 21 kilowatt (kW) system is used to treat groundwater at 150 L/min (40 gal/min); in Case 2, the same system is used to treat the groundwater at 75 L/min (20 gal/min).

Tables 5.3 and 5.4 present the costs compiled in this analysis for Case 1 and Case 2, respectively. Additional analysis is provided in these tables that compares the costs of addressing both with a 45 kW system and a 75 kW system. In Case 1, the 45 kW system treats groundwater at 300 L/min (80 gal/min), and the 75 kW unit treats it at 490 L/min (130 gal/min). In Case 2, the 45 kW system treats groundwater at 150 L/min (40 gal/min), and the 75 kW unit treats it at 250 L/min (65 gal/min).

Site-specific factors can affect the costs of using the E-beam treatment system. These factors can be divided into the following two categories: *waste-related factors* and *site features*.

Waste-related factors affecting costs include waste volume, contaminant types and levels, treatment goals, and regulatory requirements. Waste volume affects total project costs because a larger volume takes longer to remediate. However, economies of scale are realized with a larger volume project when the fixed costs, such as those for equipment, are distributed over the larger volume. The contaminant types and levels in the groundwater and the treatment goals for the site determine:

- the appropriate E-beam treatment system size, which affects capital equipment costs;
- the flow rate at which treatment goals can be met; and
- periodic sampling requirements, which affect analytical costs.

Regulatory requirements also affect permitting costs and effluent monitoring costs.

Site features affecting costs include groundwater recharge rates, groundwater chemistry, site accessibility, availability of utilities, and geographic location. Groundwater recharge rates affect the time required for cleanup. Groundwater alkalinity may increase or decrease E-beam technology REs depending on the contaminant involved. Site accessibility, availability of utilities, and site location and size all affect site preparation costs.

Table 5.3
Costs Associated with the E-Beam Technology — Case 1[a]
(Alkalinity <5mg/L as $CaCO_3$ — TCE @ 28,000 µg/L and PCE @ 11,000 µg/L)

Cost Categories	Treatment System Configurations in Kilowatts (kW)					
	21 kW (40 gal/min)		45 kW (80 gal/min)		75 kW (130 gal/min)	
	Itemized ($)	Total ($)	Itemized ($)	Total ($)	Itemized ($)	Total ($)
Site Preparation[b]		175,600		219,600		241,600
Administrative	35,000		35,000		35,000	
Treatment Area Preparation	117,600		161,600		183,600	
Treatability Study and System Design	23,000		23,000		23,000	
Permitting and Regulatory[b]		5,000		5,000		5,000
Mobilization and Startup[b]		20,000		25,000		25,000
Transportation	10,000		10,000		10,000	
Assembly and Shakedown	10,000		15,000		15,000	
Equipment[b]		842,000		1,208,000		1,432,000
Labor[c]		10,000		10,000		10,000
Supplies[c]		1,700		1,700		1,700
Disposable Personal Protective Equipment	600		600		600	
Fiber Drums	100		100		100	
Sampling Supplies	1,000		1,000		1,000	

Utilities [c]	25,700	52,600	87,500
Effluent Treatment and Disposal [c]	0	0	0
Residual Waste Shipping and Handling [c]	6,000	6,000	6,000
Analytical Services [c]	24,000	24,000	24,000
Equipment Maintenance [c]	25,300	36,200	43,000
Site Demobilization [b]	15,000	15,000	15,000
Total One-Time Costs [b]	1,057,600	1,472,600	1,718,600
Total Annual O&M Costs [c]	92,700	130,500	172,200
Groundwater Remediation			
Total Costs [d,e,f]	2,764,000	2,514,400	2,527,900
Net Present Value [g]	1,626,600	1,963,700	2,223,400
Costs per 1,000 gal [h]	5.16	6.23	7.06

[a] Costs are in February 1995 dollars
[b] Fixed costs
[c] Annual variable costs
[d] Fixed and variable costs combined
[e] Future value using annual inflation rate of 5%
[f] To complete groundwater remediation, it is assumed that the 21 kW unit will take 15 years, the 45 kW unit will take 7.5 years, and the 75 kW unit will take 4.6 years to treat 315 million gal of water.
[g] Annual discount rate of 7.5%
[h] Net present value

Source: US EPA 1995c

Table 5.4
Costs Associated with the E-Beam Technology — Case 2[a]
(Alkalinity 500 mg/L as $CaCO_3$ — Organics (See Section 5.6.9.1 Assumptions))

Cost Categories	Treatment System Configurations in Kilowatts (kW)					
	21 kW (20 gal/min)		45 kW (40 gal/min)		75 kW (65 gal/min)	
	Itemized ($)	Total ($)	Itemized ($)	Total ($)	Itemized ($)	Total ($)
Site Preparation[b]		175,600		219,600		241,600
Administrative	35,000		35,000		35,000	
Treatment Area Preparation	117,600		161,600		183,600	
Treatability Study and System Design	23,000		23,000		23,000	
Permitting and Regulatory[b]		5,000		5,000		5,000
Mobilization and Startup[b]		20,000		25,000		25,000
Transportation	10,000		10,000		10,000	
Assembly and Shakedown	10,000		15,000		15,000	
Equipment[b]		842,000		1,208,000		1,432,000
Labor[c]		10,000		10,000		10,000
Supplies[c]		1,700		1,700		1,700
Disposable Personal Protective Equipment	600		600		600	
Fiber Drums	100		100		100	
Sampling Supplies	1,000		1,000		1,000	

Utilities[c]	25,700	52,600	87,500
Effluent Treatment and Disposal[c]	0	0	0
Residual Waste Shipping and Handling[c]	6,000	6,000	6,000
Analytical Services[c]	24,000	24,000	24,000
Equipment Maintenance[c]	25,300	36,200	43,000
Site Demobilization[b]	15,000	15,000	15,000
Total One-Time Costs[b]	1,057,600	1,472,600	1,718,600
Total Annual O&M Costs[c]	92,700	130,500	172,200
Groundwater Remediation			
Total Costs[d,e,f]	6,281,600	3,994,600	3,547,200
Net Present Value[g]	2,472,900	2,350,700	2,618,100
Costs per 1,000 gal[h]	7.85	7.46	8.31

[a]Costs are in February 1995 dollars
[b]Fixed costs
[c]Annual variable costs
[d]Fixed and variable costs combined
[e]Future value using annual inflation rate of 5%
[f]To complete groundwater remediation, it is assumed that the 21 kW unit will take 30 years, the 45 kW unit will take 15 years, and the 75 kW unit will take 9.3 years to treat 315 million gal of water.
[g]Annual discount rate of 7.5%
[h]Net present value

Source: US EPA 1995c

5.6.9.1 Assumptions

The assumptions used for this analysis of E-beam technology costs are based on information provided by HVEA and observations made during the SITE demonstration.

Site-specific assumptions include the following:

- for Case 1, the contaminants and their average concentrations are TCE at 28,000 µg/L and PCE at 11,000 µg/L in groundwater which has an insignificant alkalinity of <5 mg/L as $CaCO_3$;
- for Case 2, some of the additional contaminants are saturated VOCs that are relatively difficult to treat. These VOCs are 1,1,1-TCA, 1,2-DCA, chloroform, and CCl_4; their concentrations range from 370 to 840 µg/L. The other additional contaminants are BTEX compounds present at concentrations ranging from 200 to 550 µg/L;
- the site is a Superfund site located near an urban area. As a result, utilities and other infrastructure features (for example, access roads to the site) are readily available;
- the site is located in the southeastern United States. This region has relatively mild temperatures during the winter months;
- contaminated water is located in an aquifer no more than 100 ft below ground surface; and
- the groundwater remediation project involves a total of 1,200 million L (315 million gal) of water that needs to be treated. This groundwater volume corresponds to the volume treated by a 21 kW unit operating continuously for 15 years at a flow rate of 150 L/min (40 gal/min).

Equipment assumptions include the use of the 21 kW unit treating contaminated groundwater at a rate of 150 L/min (40 gal/min) in Case 1 and 75 L/min (20 gal/min) in Case 2. The system is operated on a continuous flow cycle, 24 hours per day, 7 days per week. The system can, therefore, treat nearly 79 million L/yr (21 million gal/yr) in Case 1, and about 40 million L/yr (10.5 million gal/yr) in Case 2. Because most groundwater remediation projects are long-term projects, about 1,200 million L (315 million gal) of water are assumed to be treated in both cases. Case 1 remediation will take about 15 years to complete, and Case 2 about 30 years. In practice, it is difficult to determine both the volume of groundwater to treat and the actual duration of a project.

Neither depreciation nor salvage value is applied to the costs presented because the equipment is not purchased by a customer. All depreciation and salvage value is assumed to be incurred by the vendor and is reflected in the ultimate cost of leasing the E-beam treatment equipment.

Operating parameter assumptions using a 21 kW system are listed below:

- costs for 45 kW and 75 kW systems are presented in Tables 5.3 and 5.4;
- the treatment system is operated 24 hours per day, 7 days per week, 52 weeks per year;
- the treatment system operating at full power has a maximum voltage of 500 kV and a maximum beam current of 42 mA.
- the treatment system operates automatically without the constant attention of an operator and will shut down in the event of a malfunction;
- modular components consisting of the equipment needed to meet treatment goals are mobilized to the site and assembled by the contractor;
- air emissions monitoring is not necessary; and
- E-beam equipment will be maintained by the contractor and will last for the duration of the groundwater remediation project with proper maintenance.

Total costs are presented as future values based on the following financial conditions. The costs per 1,000 gal (3,800 L) treated are presented as net present values and assumes a 5% annual inflation rate to estimate the future values. The future values are presented as net present values using a discount rate of 7.5% (using a higher discount rate makes the initial costs weigh more heavily). Because the costs of demobilization will occur at the end of the project, the appropriate future values of these costs were used to calculate the totals at the bottom of Tables 5.3 and 5.4.

Additional assumptions include:

- costs are rounded to the nearest $100;
- contaminated groundwater is treated to achieve the removal efficiencies (REs) observed in SITE demonstration Runs 3 and 13 for Cases 1 and 2, respectively;
- the E-beam system is mobilized to the remediation site from within 500 miles of the site;
- operating and sampling labor costs are incurred by the client. The vendor performs maintenance and modification activities that are paid for by the client;
- initial operator training is provided by the vendor; and
- four groundwater extraction wells already exist on-site. They are assumed to be capable of providing the flow rates discussed in this economic analysis.

Cost data are presented for the following categories:

- site preparation;
- permitting and regulatory;
- mobilization and start-up;
- equipment;
- labor;
- supplies;
- utilities;
- effluent treatment and disposal;
- analytical services;
- equipment maintenance; and
- site demobilization.

Each of these categories is discussed below.

5.6.9.2 Site Preparation Costs

Site preparation costs include administrative, treatment area preparation, treatability study, and system design costs. For this analysis, site preparation and administrative costs, such as those for legal searches, access rights, and site planning activities, are estimated to be \$35,000.

Treatment area preparation includes constructing a shelter building and installing pumps, valves, and piping from the extraction wells to the shelter building. The shelter building needs to be constructed before mobilization of the E-beam system. A 37 m^2 (400 ft^2) building is required for the 21 kW system. The 45 kW system requires 74 m^2 (800 ft^2) of building space, and the 75 kW system requires 92 m^2 (1,000 ft^2). Construction costs are estimated to be \$1,184/m^2 (\$110/ft^2), which covers installation of radiation shielding materials. A natural gas heating and cooling unit and related ductwork is estimated at \$20,000 installed. The total shelter building construction costs for the 21 kW system are estimated to be \$64,000.

Four extraction wells are assumed to exist on-site which are located 61 m (200 ft) from the shelter building. Four 132 L/min (35 gal/min), 1.5 hp, variable-speed Teflon® pumps are required to maintain the flow rates necessary for each case. The total installed cost for the pumps, including electrical equipment, is \$5,600. Piping and valve connection costs are \$20/m (\$6/ft), which covers underground installation. The total piping cost is \$48,000. Thus, total site preparation cost is estimated to be \$117,600.

HVEA estimates the treatability study to cost \$18,000, including labor and equipment costs. System design includes determining which E-beam system will achieve treatment goals and designing the configuration. The system design is estimated to cost \$5,000.

5.6.9.3 Permitting and Regulatory Costs

Permitting and regulatory costs in this analysis include permit fees for discharging treated water to a surface water body. The cost of this permit is based on regulatory agency requirements and treatment goals for a particular site. The discharge permit for each case is estimated to cost \$5,000. Costs of highway permits for overweight vehicles are included in the costs of mobilization.

5.6.9.4 Mobilization and Start-up Costs

Mobilization and start-up costs include the costs of transporting the E-beam system to the site, assembling the E-beam system, and performing the initial shakedown of the treatment system. HVEA provides initial operator training to its clients as part of providing the E-beam equipment.

Transportation costs are assumed at $6.21/km ($10/mi) for 621 km (1,000 mi), or $10,000. The costs of highway permits for overweight vehicles are included in this total cost.

Assembly costs include the costs of unloading equipment from the trailers, assembling the E-beam system, and connecting extraction well piping and electrical lines. A two-person crew will work three 8-hour days to unload and assemble the system and perform the initial shakedown. The total start-up costs are assumed as $10,000, including labor and hookup costs.

For the 45 kW and 75 kW systems, completion of initial assembly and shakedown activities is expected to require the two-person crew to work about five 8-hour days. The start-up costs for these systems are about $15,000, including labor and electrical hookup costs. Total mobilization and start-up costs for each case are estimated to be about $20,000.

5.6.9.5 Equipment Costs

Equipment costs include the costs of leasing the E-beam treatment system. HVEA provides the complete E-beam treatment system configured for site-specific conditions. All E-beam treatment equipment is leased to the client. As a result, all depreciation and salvage value is reflected in the price for leasing the equipment. At the end of a treatment project, HVEA decontaminates and demobilizes its treatment equipment.

Equipment costs are determined by the size of the E-beam system needed to complete the remediation project and are incurred as a lump sum; as a result, even though the equipment is leased to the client, it is not priced at a monthly rate. For this analysis, HVEA estimates that the capital equipment for both cases will cost $842,000 for a 21 kW system; $1,208,000 for a 45 kW system; and $1,432,000 for a 75 kW system.

5.6.9.6 Labor Costs

Once the system is functioning, it is assumed to operate continuously at the design flow rate except during routine maintenance, which HVEA conducts. One operator performs routine equipment monitoring and sampling activities. Under normal operating conditions, an operator is required to monitor the system about once each week.

It is also assumed that system monitoring and sampling duties is conducted by a full-time employee of the site owner who is assigned as the primary operator. Further, a second person, also employed by the site owner, will be trained to act as a backup operator. Based on observations made at the SITE demonstration, it is estimated that operation of the system requires about one-quarter of the primary operator's time. Assuming the primary operator earns $40,000/year, the total direct annual labor cost for each case is estimated to be $10,000.

5.6.9.7 Supply Costs

No chemicals or treatment additives are typically used to treat the groundwater using E-beam technology. Therefore, no direct supply costs are expected. Supplies that will be needed as part of the overall groundwater remediation project include Level D, disposable personal protective equipment (PPE), PPE disposal drums, and sampling and field analytical supplies.

Disposable PPE for each case is assumed to cost about $600/yr for the primary operator. Used PPE is assumed to be hazardous and needs to be disposed of in 90 L (24 gal) fiber drums. One drum is assumed to be filled every 2 months, and each drum costs about $12. For each case, the total annual drum costs are about $100.

During the demonstration at Savannah River Site, the average pH level of the influent was about 4.7; the average pH level of the effluent ranged between 3.0 and 3.5. Depending on discharge permit levels and influent and effluent pH levels, the pH may require adjustment. In this event, additional supplies will be necessary. The quantity of supplies needed is highly site-specific and difficult to determine; therefore, this analysis does not include posttreatment pH adjustment costs.

Total annual supply costs for each case are estimated to be $1,700.

5.6.9.8 Utility Costs

Electricity is the only utility used by the E-beam system. Electricity is used to run the E-beam treatment system, pumps, blower, and air chiller. Electricity costs can vary considerably depending on the location of the site, local utility rates, the E-beam system used, the total number of pumps and other electrical equipment operating, the use of the air chiller, and whether electrical power lines are available at the site or must be installed.

This analysis assumes that power lines are available at the site, and a constant rate of electricity consumption based on the electrical requirements of the 21 kW E-beam treatment system. The pumps, blower, and air chiller are assumed to draw an additional 20 kW, which is based on observations made during the SITE demonstration at the Savannah River Site. Therefore, the 21 kW unit operating for 1 hour draws about 42 kW hours (kWh) of electricity. The total annual electrical energy consumption is estimated to be about 366,910 kWh. Electricity is assumed to cost $0.07/kWh, including demand and usage charges. The total annual electricity costs for each case are estimated to be about $25,700. The total annual electricity costs are estimated to be $52,600 for the 45 kW system and $87,500 for the 75 kW system.

Water and natural gas usage are highly site-specific, but assumed to be minimal for each case in this analysis. As a result, no costs for these utilities are included.

5.6.9.9 Effluent Treatment and Disposal Costs

At the Savannah River Site demonstration, the E-beam system did not meet target treatment levels for about half of the VOCs. Depending on the treatment goals for a site, additional effluent treatment may be required and additional treatment or disposal costs incurred. Because of this uncertainty, effluent treatment or disposal costs are not included.

The E-beam system does not produce air emissions because the water delivery and cooling air systems are enclosed. As a result, no cost for air emissions treatment is incurred.

5.6.9.10 Residual Waste Shipping and Handling Costs

The only residuals produced during E-beam system operation are fiber drums containing used PPE and waste sampling and field analytical supplies, all of which are typically associated with a groundwater remediation project. This waste is considered hazardous and requires disposal at a permitted facility. For each case, it is assumed that about six drums of waste are disposed annually. The cost of handling and transporting the drums and disposing them at a hazardous waste disposal facility is about $1,000 per drum. The total drum disposal costs for each case are about $6,000/yr.

5.6.9.11 Analytical Services Costs

Required sampling frequencies and number of samples are highly site-specific and are based on treatment goals and contaminant concentrations. Analytical costs associated with a groundwater remediation project include the costs of laboratory analyses, data reduction, and Quality Assurance/Quality Control (QA/QC). The analysis assumes that one sample of untreated water, one sample of treated water, and associated QC samples (trip blanks, field duplicates, and matrix spike/matrix spike triplicates) will be analyzed for VOCs every month. Therefore, monthly analytical costs are about $2,000.

5.6.9.12 Equipment Maintenance Costs

HVEA estimates that annual equipment maintenance costs are about 3% of the capital equipment costs. Therefore, the total annual equipment maintenance costs for each case are about $25,300 for the 21 kW system, $36,200 for the 45 kW system, and $43,000 for the 75 kW system.

5.6.9.13 Site Demobilization Costs

Site demobilization includes treatment system shut-down, disassembly, and decontamination; site cleanup and restoration; utility disconnection; and transportation of the E-beam equipment off-site. A two-person crew will work about five 8-hour days to disassemble and load the system. It is assumed that the equipment will be transported 1000 km (670 mi) either for storage or to the next job site. HVEA estimates that the total cost of demobilization is about $15,000 for each case. This total includes all labor, material, and transportation costs.

5.6.9.14 Economic Analysis Conclusions

Total estimated fixed costs are about $1,057,600 for each case. Of this total, $842,000, or about 80%, is for E-beam equipment costs. Over 16% of the total fixed cost is for site preparation; this cost is not entirely attributable to operating the treatment system, but rather is necessary for setting up the system. Total estimated annual variable costs are about $92,700 for each case. Of this total, analytical service costs comprise about 26%, equipment maintenance costs about 27%, and utility costs nearly 28%.

The analysis of the base-case E-beam technology (21 kW) reveals that operating costs are strongly affected by the E-beam system and flow rate used. The larger systems take less time to complete a groundwater remediation project, but the higher equipment and utility costs result in a higher cost per 3,785 L (1,000 gal) of groundwater treated. The base-case assumes that the total amount of groundwater to be treated is 1,200 million L (315 million gal). In Case 1, 15 years would be needed to complete the remediation project; in Case 2, 30 years would be needed. The total estimated cost of the project is $2,764,000 for Case 1 and $6,281,000 for Case 2. The estimated cost per 3,785 L (1,000 gal) of groundwater treated (net present value) is $5.16 for Case 1 and $7.85 for Case 2.

Table 5.5 presents only the direct costs associated with the E-beam treatment system. This analysis is provided to segregate the direct costs of procuring and operating the E-beam system from the total costs of a groundwater remediation project. The direct costs are the same for both cases. Total fixed costs are estimated to be $900,000, and total annual variable costs are estimated to be $67,000. The analytical supplies cost has been excluded because at $1,000/yr, it represents about 1% of the total annual variable costs. The direct cost per 3,785 L (1,000 gal) of groundwater treated is estimated to be $4.07 for Case 1 and $5.99 for Case 2.

In summary, the cost of treatment using an HVEA E-beam system depends on may factors such as the initial concentrations of organic contaminants, treatment objectives, the dose required to obtain the desired destruction, the volume of waste to be treated, the size of the treatment facility, the length of treatment, and the manner in which capital recovery is handled. The cost of treatment using HVEA systems in various industrial waste and groundwater applications has ranged from $52/1,000 L to $7.57/L ($2/1,000 gal to $0.50/gal).

5.6.10 Design Validation

Validation of the design of an E-Beam treatment system is accomplished using treatability studies to establish the necessary electron beam energy required and to ascertain the nature and extent of posttreatment, if any, that is required. The simplicity of the system enables these studies to be accomplished economically.

Table 5.5
E-Beam Treatment System Direct Costs[a]

Cost Categories	Treatment System Configurations in Kilowatts (kW)					
	21 kW		45 kW		75 kW	
	Itemized ($)	Total ($)	Itemized ($)	Total ($)	Itemized ($)	Total ($)
Site Preparation[b]		23,000		23,000		23,000
Treatability Study and System Design	23,000		23,000		23,000	
Mobilization and Startup[b]		20,000		25,000		25,000
Transportation	10,000		10,000		10,000	
Assembly and Shakedown	10,000		10,000		10,000	
Equipment[b]		842,000		1,208,000		1,432,000
Labor[c]		10,000		10,000		10,000
Utilities[c]		25,700		52,600		87,500
Residual Waste Shipping and Handling[c]		6,000		6,000		6,000
Equipment Maintenance[c]		25,300		36,200		43,000
Site Demobilization[b]		15,000		15,000		15,000
Total One-Time Costs[b]		900,000		1,271,000		1,495,000
Total Annual O&M Costs[c]		67,000		104,800		146,500
Cost per 1,000 gal Treated — Case 1[d]		4.07		5.17		6.07
Costs per 1,000 gal Treated — Case 2[e]		5.99		6.05		7.10

[a]This table presents direct costs associated with the E-beam treatment system segregated from the costs incurred as a result of conducting a groundwater remediation project. All assumptions used in this analysis apply.
[b]Fixed costs
[c]Variable costs
[d]Net present value using the same assumptions used in Table 5.3
[e]Net present value using the same assumptions used in Table 5.4

Source: US EPA 1995c

5.6.11 Permitting Requirements

The process will most likely require acquisition of one or more discharge permits for the treated water. No air permits are required. In addition, the electron beam source requires that the site of operation have a permit for the use of an ionizing radiation source. This permit is usually issued by the health department for the county or the state and is the same permit as required for x-ray equipment.

5.6.12 Performance Measures

The system performance is determined by sampling and analyzing the quality of the treated water.

5.6.13 Design Checklist

Following is a list of elements to be considered during design.

1. Quality and flow rate of the stream to be treated.
2. Characteristics of stream to be treated: pH, carbonates, concentrations of organics, and dissolved solids.
3. Target limits for treated effluent BOD, COD, etc.
4. Site utilities: electricity, water (drinking, sanitary, process), and telephone.
5. Availability of support services, such as fire fighting, emergency medical, etc.
6. Site accessibility for equipment delivery by road or rail, restrictions on loading, noise, etc., and availability of access for system operation and maintenance. Maintenance access roads.

5.7 Implementation and Operation

5.7.1 Implementation Strategies

The E-beam and its design are well developed; however, its implementation is highly site-specific. Implementation must begin with a good understanding of the site and the contaminants. This information should be coupled with a set of treatability studies performed on samples of the groundwater from the site using equipment on which data has been obtained and on which successful scale-up have occurred at other sites. Conducting the treatability study on actual site samples is especially crucial for this technology since small differences in chemical composition can have a major impact on the performance.

The vendor used for these treatability studies can be the supplier of the equipment or an independent party; however, a strong background in the use of electron beam treatment is necessary. The vendor should be capable of interpreting the results of the treatability studies and creating a set of performance specifications for the pieces of equipment. The equipment itself is covered by patents held by the developer and, in general, it can only be acquired through High Voltage Environmental Applications, Inc. While competitive bidding is desirable, one must recognize that this application of high voltage at a large scale and relatively high power consumption under highly humid conditions requires extensive field experience and the ultimate vendor selection process should include assessment of the vendor's experience in similar applications in addition to a cost comparison.

5.7.2 Operation

Startup of the full-scale system was observed by the author of this chapter during a visit in April 1996. Startup consisted of clearing the irradiation chamber of personnel, slowly warming up the power supplies for the electron beam by increasing the power level to the operating power over approximately 15 to 20 minutes, then starting the flow of the contaminated stream. The absence of moving parts is apparent.

Operation involves checking and adjusting pH and alkalinity of the feedwater, periodic checks of the electric systems, and sampling and analysis of the influent and effluent from the process. Once operating, the system requires only a part-time person to maintain and monitor its performance.

Successful installation requires that the pH and alkalinity of the influent be maintained within design specifications. The pH monitoring and automatic adjustment using acid or alkaline solutions is a highly desirable addition to the system. Otherwise, the system requires minimal monitoring.

5.8 Case History

The E-beam technology was extensively evaluated by the Superfund Innovative Technology Evaluation (SITE) program of the U.S. Environmental Protection Agency (US EPA), Risk Reduction Engineering Laboratory, now named National Risk Management Research Laboratory (NRML), Cincinnati, Ohio. The report (US EPA 1995d) presents detailed performance and cost information for the process as well as discussing the case history of the demonstration program. The demonstration was conducted at the U.S. Department of Energy (DOE) Savannah River Site (SRS) in Aiken, South Carolina, during two different periods totaling 3 weeks in September and November 1994.

During the demonstration, about 265,000 L (70,000 gal) of M-area groundwater contaminated with VOCs was treated. The principal groundwater contaminants were trichlorethylene (TCE) and perchloroethylene (PCE), which were present at concentrations of about 27,000 and 11,000 μg/L, respectively. The groundwater also contained low levels (40 μg/L) of cis-1,2-dichloroethene (1,2-DCE). Before treatment, groundwater was pumped from a recovery well into a 28,000 L (7,500 gal) equalization tank to minimize any variability in influent characteristics. Treated groundwater was stored in a 38,000 L (10,000 gal) tank before being pumped to an on-site air stripper, which was treating contaminated groundwater from the demonstration area.

During a portion of the E-beam technology demonstration, the groundwater was spiked with VOCs not present in the M-area groundwater. The resultant influent concentrations ranged from about 100 to 500 μg/L for the following spiking compounds: 1,1,1-trichloroethane (1,1,1-TCA), 1,2-dichloroethane (1,2-DCA); chloroform; carbon tetrachloride (CCl_4); and benzene, toluene, ethylbenzene, and p-xylene (BTEX). Saturated VOCs (1,1,1-TCA, 1,2-DCA, chloroform, and CCl_4) were chosen as spiking compounds because they are relatively difficult to destroy using technologies such as the E-beam technology that involve free radical chemistry. BTEX were chosen because they are common groundwater contaminants at Superfund and other contaminated sites. For the SITE technology demonstration, TCE, PCE, 1,2-DCE, and the spiking compounds were considered to be critical VOCs.

5.8.1 Demonstration Procedures

The technology demonstration was conducted in five phases. Thirteen test runs were performed during these five phases to evaluate HVEA treatment system performance. During each run, influent characteristics or operating parameters were changed to collect information in order to meet project objectives. The demonstration approach is summarized below.

During Phase 1, beam current, one of the principal operating parameters, was varied to observe how E-beam dose affects treatment system performance at a constant flow rate of 150 L/min (40 gal/min). Three runs were conducted during Phase I using unspiked groundwater.

During Phase 2, spiked groundwater was used to collect information on treatment system performance in destroying VOCs other than those present in the M-area groundwater. Two runs were performed using different beam currents and a constant flow rate of 150 L/min (40 gal/min). Phase 2 also included a zero dose run to identify reduction in VOC concentrations resulting from mechanisms other than VOC destruction by the E-beam (for example, volatilization).

Phase 3 tested the reproducibility of HVEA system performance for treating spiked groundwater. Three runs were performed under identical operating conditions, which were determined based on preliminary treatment results from Phases 1 and 2.

Phase 4 consisted of one run to evaluate HVEA system performance at the minimum limiting flow rate [57 L/min (15 gal/min)] of the system used for the demonstration. The minimum flow rate was chosen because preliminary results from Phases 1, 2, and 3 indicated that the HVEA system did not meet effluent target levels at higher flow rates and using maximum beam current.

Phase 5 began 4 weeks after Phase 4 was completed. The interval between Phases 4 and 5 gave HVEA time to evaluate preliminary results from Phases 1 through 4 and conduct additional studies on spiked and unspiked groundwater from M-area recovery well RWM-1 in order to determine Phase 5 operating conditions. Based on information from the test runs and additional studies, HVEA adjusted the influent delivery system to improve overall treatment system performance. This was accomplished by increasing the dose without increasing the beam current or lowering the flow rate.

To evaluate the effect of the improved delivery system in Phase 5, HVEA selected the same flow rate (75.71 L/min [20 gal/min]) and beam current (42 mA [63 kW]) as used in the reproducibility runs. Of the three Phase 5 runs, one used unspiked groundwater, one used spiked groundwater, and one used alkalinity-adjusted spiked groundwater. Alkalinity was adjusted in one run because carbonate and bicarbonate ions scavenged •OH, potentially affecting VOC removal efficiency. During this run, a sodium bicarbonate solution was added to the influent in order to adjust alkalinity from <5 mg/L to about 500 mg/L as $CaCO_3$, which is within the typical range of groundwater alkalinity levels in the United States.

5.8.2 Sampling and Analytical Procedures

During the demonstration, groundwater samples were collected at E-beam influent and effluent sampling locations, and cooling air samples were collected before and after the carbon adsorber.

Each test run lasted about 3 hours, and four groundwater sampling events were conducted at 45-minute intervals during each test run. Groundwater samples for VOC analysis were collected during each sampling event so that average influent and effluent concentrations could be calculated based on four replicate data points. Groundwater samples for other analyses were typically collected during two of the sampling events.

Groundwater samples were collected during all runs for VOC and pH analyses. Groundwater samples were collected during selected runs for analysis for SVOCs, haloacetic acids, aldehydes, H_2O_2 (effluent only), TOC, purgeable organic carbon (POC), total inorganic carbon (TIC), total organic halides (TOX), chloride, alkalinity, and acute toxicity. Influent and effluent samples were analyzed using US EPA-approved methods, such as those found in *Test Methods for Evaluation Solid Waste and Methods for Chemical Analysis of Water and Wastes* (US EPA 1990; US EPA 1983) or other standard or published methods (American Public Health Association, American Water Works Association, and Water Environment Federation 1992; Boltz and Howell 1979).

During Runs 1 through 10, cooling air samples were collected and analyzed for VOCs, O_3, and HCl using an on-site Fourier transform infrared (FTIR) interferometer. Cooling air samples were not collected during Runs 11, 12, and 13 (Phase 5) because of high costs associated with maintaining the FTIR interferometer in the field during the 4-week interval between Phases 4 and 5. This approach did not affect project objectives because cooling air was analyzed only for noncritical parameters to meet a secondary objective.

On-site measurements of flow rate, beam current, and power consumption were recorded during all runs.

In all cases, US EPA-approved sampling, analytical, and QA/QC procedures were followed to obtain reliable data. These procedures are described in the QAPP written specifically for the E-beam technology demonstration (PRC Environmental Management 1994) and are summarized in the TER, which is available from the US EPA project manager.

5.8.3 Removal Efficiency

Table 5.6 presents the range of critical VOC concentrations in the influent to the E-beam unit for unspiked and spiked test runs.

Most of the performance data are reported based on average values from replicate sampling events. In some cases, samples were analyzed at two dilutions; when this occurred, the results for the lower dilution were used to calculate the average value. For influent samples with analyte concentrations at nondetectable levels, the detection limit was used as the estimated concentration when the average value was calculated. For effluent samples with analyte concentrations at nondetectable levels, one-half the detection limit was used as the estimated concentration when the average value was calculated. If all replicate effluent samples had nondetectable concentrations of any analyte, the detection limit was used as the average value, the removal efficiency was reported as a greater than (>) value, and the 95% upper confidence limit (UCL) was not calculated.

After the demonstration data were reviewed, it was determined that more than one approach should have been used to handle nondetectable influent and effluent values in order to calculate averages. For influent nondetectable values, the detection limit was used in place of the nondetectable value. Although this approach deviates from typical environmental engineering practice, which is to use one-half the detection limit for nondetectable values, using the full detection limit is more appropriate in this case because (1) there is less variability in influent concentrations than in effluent concentrations, and (2) 50% or more of the influent samples had VOC concentrations above the detection limit. For example, in about 50% of the influent samples (28 out of 52 samples collected in 13 runs), the concentration of 1,2-DCE was reported as nondetectable (the detection limit is 40 µg/L); in the remaining samples, this compound was present at concentrations of up to 50 µg/L. For effluent nondetectable values, however, using the typical practice for handling nondetectable values is more appropriate. This is the case because the effluent data have greater variability as a result of E-beam treatment and because the data are more limited (the effluent data from all the runs cannot be pooled together because the operating conditions generally varied from run to run).

Table 5.6
VOC Concentrations in Unspiked and Spiked Groundwater Influent

VOC	Unspiked Groundwater (µg/L)	Spiked Groundwater (µg/L)
TCE	25,000 to 30,000	25,000 to 37,000
PCE	9,200 to 12,250	9,200 to 14,000
1,2-DCE	40 to 43	<40 to 45
1,1,1-TCA	ND[a]	200 to 500
1,2-DCA	ND[b]	210 to 840
Chloroform	ND[a]	240 to 650
CCl_4	ND[a]	150 to 400
Benzene	ND[a]	220 to 550
Toluene	ND[a]	170 to 360
Ethylbenzene	ND[a]	95 to 250
Xylenes	ND[a]	85 to 200

ND[a] Not detected, detection limit not given, estimated to be about 1 µg/L
ND[b] Not detected, detection limit 40 µg/L

Source: US EPA 1995c

Table 5.7 summarizes the VOC removal efficiencies for unspiked and spiked groundwater runs conducted at different E-beam doses. HVEA controls dose by adjusting the beam current, the flow rate, and the thickness of the water stream impacted by the E-beam. In HVEA's system, the beam current is controlled directly from the control panel, the flow rate is controlled by manually adjusting the influent pump, and the thickness of the water stream is controlled by the influent delivery system.

During Phase 1, unspiked groundwater was treated at three different doses. For these runs, the dose was varied by changing the beam current, while the flow rate remained constant. As shown in Table 5.7, removal efficiencies for TCE, PCE, and 1,2-DCE increased when the beam current was increased. A similar effect was observed during Phase 2, which involved spiked groundwater. The dose was increased further during Phase 3 by lowering the flow rate from 150 to 75 L/min (40 to 20 gal/min) and increasing the beam current to the maximum level (42 mA [63 kW]); corresponding increases in REs were observed, particularly for spiked compounds. Finally, for Phase 5, HVEA adjusted the delivery system and these adjustments increased the dose although the beam current and flow rate were set at the same levels as were used for Phase 3. HVEA considers information regarding the delivery system to be proprietary. Phase 5 results indicate that the delivery system adjustments increased removal efficiencies for most VOCs. In fact, the operating conditions during Phase 5 generally yielded the highest removal efficiencies observed during the demonstration.

Table 5.7 also shows that for all spiked groundwater runs, removal efficiencies for TCE, PCE, 1,2-DCE, and BTEX were much higher than for 1,1,1-TCA, 1,2-DCA, chloroform, and CCl_4. The difference in system performance for these two groups of VOCs is postulated to be due to the presence of double bonds between carbon atoms in TCE, PCE, and 1,2-DCE and aromatic bonds between carbon atoms in BTEX, which makes these compounds more amenable to oxidation by free radicals generated by the E-beam. Furthermore, the removal efficiencies of saturated chlorinated compounds, typified by CCl_4, were consistently higher than removal efficiencies for 1,1,1-TCA, 1,2-DCA, and chloroform. This effect may be a consequence of the relatively large number of chlorine atoms in CCl_4. The four chlorine atoms facilitate CCl_4 destabilization and are good "leaving groups" in the presence of free radicals; therefore, CCl_4 may be more amenable to E-beam destruction than similar compounds with fewer chlorine atoms.

Table 5.7
VOC Removal Efficiencies (REs)

VOC	VOC RE (%)							
	Phase 1			Phase 2		Phase 3	Phase 5	
	Run 1 BC = 7 FR = 40	Run 2 BC = 14 FR = 40	Run 3 BC = 42 FR = 40	Run 4 BC = 17 FR = 40	Run 5 BC = 21 FR = 40	Runs 7,8, and 9 BC = 42 FR = 20	Run 11 BC = 42 FR = 20	Run 12 BC = 42 FR = 20
TCE	73	92	97	91	93	94 to 96	>99	98
PCE	50	83	96	76	87	95	99	99
1,2-DCE	>53	>83	>90	NA	>85	85 to >91	>91	>88
1,1,1-TCA				17	33	61 to 62		73
1,2-DCA				25	30	60 to 65		70
Chloroform				24	28	56 to 57		68
CCl_4				46	73	89 to 91		>97
Benzene				96	98	93 to 97		>98
Toluene				>97	>96	95 to 96		>98
Ethylbenzene				>93	>94	95 to 97		>97
Xylenes				>91	>93	>93 to 95		>96

BC beam current (mA)
FR flow rate (gal/min)
NA not applicable (because 1,2-DCE was not detected in any sample collected during Run 4)

Source: US EPA 1995c

Table 5.8 shows that the HVEA treatment system achieved the effluent target levels for 1,2-DCE, CCl_4, and BTEX. Effluent target levels were not achieved for 1,1,1-TCA, 1,2-DCA, and chloroform when these compounds were present at spiked levels (230,440, and 316 µg/L for 1,1,1-TCA, 1,2-DCA, and chloroform, respectively). Effluent target levels were also not achieved for TCE and PCE when they were present at existing levels in M-area groundwater (27,000 and 11,000 µg/L for TCE and PCE, respectively).

Only effluent concentrations for Runs 11 and 12 are shown in Table 5.8 because the HVEA treatment system displayed the best overall performance, in terms of removal efficiencies during these runs. However, effluent target levels were met for toluene, ethylbenzene, and xylenes during the other runs.

Table 5.8

Compliance with Applicable Effluent Target Levels

VOC	Effluent Target Level (µg/L)	95 Percent UCL for Effluent Concentration (µg/L)	
		Run 11	Run 12
TCE	5	190	1,100
PCE	5	100	250
1,2-DCE[a]	54	4U	4U
1,1,1-TCA	54	NA	83
1,2-DCA	5	NA	180
Chloroform	46	NA	130
CCl_4	5	NA	4U
Benzene	5	NA	4U
Toluene	80	NA	4U
Ethylbenzene	57	NA	4U
Xylenes[a]	320	NA	4U

U analyte not detected in the treatment system effluent at or above the value shown
NA not applicable (because the analyte was not detected in the treatment system influent)

[a] Influent concentrations for 1,2-DCE and xylenes were below the effluent target levels

Source: US EPA 1995c

5.8.4 Effect of Treatment on Toxicity

Bioassay tests were performed to evaluate the change in acute toxicity of the groundwater after treatment by the HVEA system. Two common freshwater test organisms, a water flea (*Ceriodaphnia dubia*) and a fathead minnow (*Pimephales promelas*), were used in the bioassay tests. The acute toxicity was measured as the concentration at which 50% of the organisms died (LC50) and was expressed as the % age of influent or effluent in the test water. One influent sample and one effluent sample from each run were tested; chronic toxicity was not measured.

Table 5.9 presents the bioassay test results of influent and effluent samples from Runs 4 and 5 (the reproducibility runs) and Runs 11, 12, and 13. These results show that some influent samples and all effluent samples were acutely toxic to both test organisms. The change in groundwater toxicity resulting from treatment by the HVEA system was evaluated statistically using data from the reproducibility runs. Specifically, the mean difference between the influent and effluent LC50 values was compared to zero using a two-tailed paired Student's t-test. The null hypothesis was that the mean difference between influent and effluent LC50 values equaled zero at a 0.05 significance level. The critical t value at this significance level with two degrees of freedom is 4.303. The calculated t values for the water flea and the fathead minnow were 1.47 and 31.6, respectively. These results indicate that treatment by the E-beam technology statistically increased groundwater toxicity for the fathead minnow, but not for the water flea.

As noted above, influent and effluent samples for bioassay testing were collected during Runs 11, 12, and 13, which were conducted after HVEA adjusted the influent delivery system to increase the dose. Although toxicity data for these runs cannot be statistically evaluated because the influent characteristics were different, the data suggest that the difference between the influent and effluent LC50 values decreased for fathead minnows. For fathead minnows, the increase in toxicity resulting from E-beam treatment (the difference between influent and effluent LC50 values) in Run 12 was less than the average increase in toxicity in Runs 8 and 9. This fact may be related to the higher VOC REs and reduced byproduct formation achieved when the dose was increased by adjusting the influent delivery system for Run 12. However, for water fleas, the LC50 data could not be compared because the increases in LC50 values resulting from E-beam treatment were not observed to be absolute values (that is, all observations were > values).

Table 5.9
Acute Toxicity Data

Run[a]	LC50 (%)			
	Ceriodaphnia dubia		Pimephales promelas	
	Influent	Effluent	Influent	Effluent
4	35	8.8	72	8.6
5	68	18	100	15
7[b]	>100	17	>100	8.8
8[b]	17	<6.2	>100	16
9[b]	16	<6.2	>100	18
11	37	<6.2	89	8.8
12	37	<6.2	83	9.8
13	>100	12	>100	54

[a]Runs 7, 8, and 9 were the reproducibility runs (Phase 3). Runs 11, 12, and 13 were conducted after HVEA adjusted the influent delivery system (Phase 5). Run 11 was conducted with unspiked groundwater, Run 12 was conducted with spiked groundwater, and Run 13 was conducted with alkalinity-adjusted spiked groundwater.
[b]Using data from the three reproducibility runs, a two-tailed paired Student's t-test with a 0.05 significance level was performed for each organism. The null hypothesis was that the mean difference between the influent and effluent LC50 values equaled zero. For LC50 values shown as >100 and <6.2, 100 and 6.2 were used to calculate the mean difference. The calculated t values were 1.47 and 31.6 for Ceriodaphnia dubia and Pimephales promelas, respectively.

Source: US EPA 1995c

Published reports indicate the H_2O_2 generated by technologies involving free radicals may contribute to effluent toxicity (US EPA 1993). The average effluent H_2O_2 concentration was 8.0 mg/L during the reproducibility runs and 8.9 mg/L during Phase 5 runs. Literature data indicate that the LC50 for H_2O_2 for the water flea is about 2 mg/L. Because no statistically significant increase in acute toxicity for the water flea was observed during E-beam treatment despite high levels of H_2O_2 in the effluent, it is likely that any increase in toxicity associated with H_2O_2 was counteracted by a decrease in toxicity resulting from VOC removal. The fathead minnow is less sensitive to H_2O_2 than the water flea. The Connecticut Department of Environmental Protection (CDEP) reported an LC50 value of 18.2 mg/L of H_2O_2 with 95% confidence limits of 10 and 25 mg/L for the fathead minnow (CDEP 1993). Therefore, the statistically significant increase in acute toxicity for the fathead minnow during E-beam treatment is more likely to have been caused by residual VOCs or treatment byproducts than by H_2O_2.

Figure 5.2
VOC REs in Reproducibility Runs

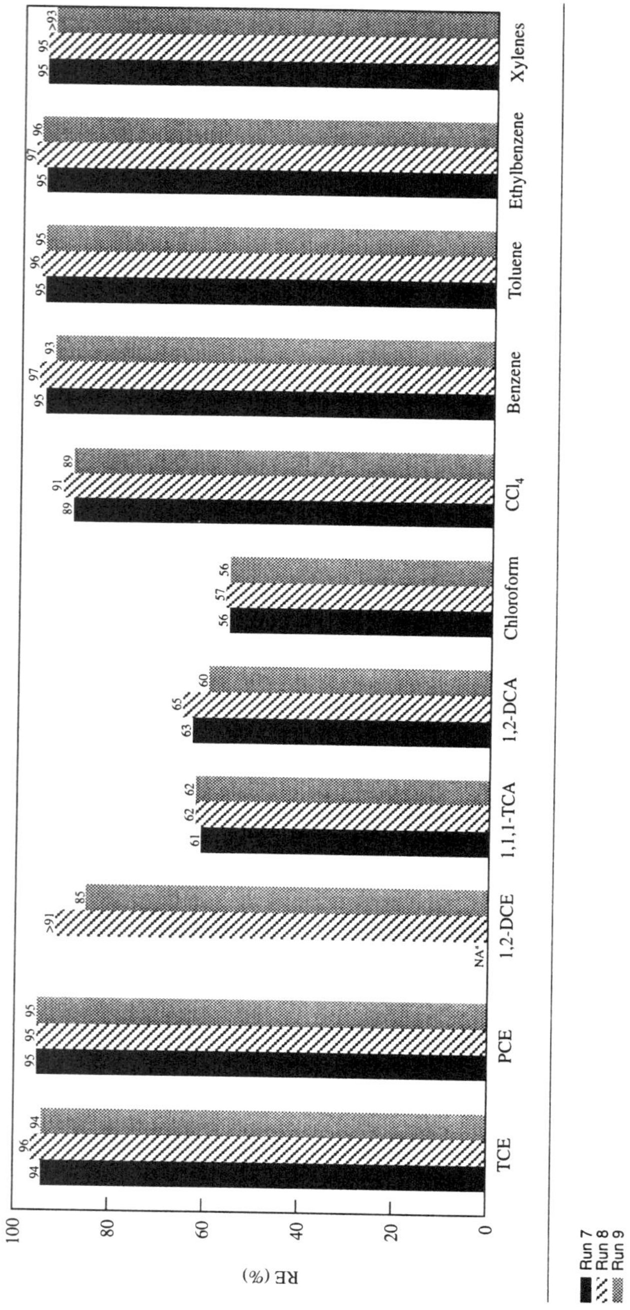

[a]NA = 1,2-DCE was not detected in Run 7

Reproduced courtesy of High Voltage Environmental Applications, Inc.

5.8.5 Reproducibility of Treatment System Performance

VOC removal efficiencies in the Phase 3 reproducibility runs (Runs 7, 8, and 9) are shown in Figure 5.2. This figure indicates that the removal efficiencies for all VOCs were reproducible. The maximum difference among removal efficiencies for the three runs occurred for 1,2-DCA, for which REs ranged from 60 to 65%, and 1,2-DCE, for which removal efficiencies ranged from 85 to >91%. However, for other VOCs, the removal efficiencies differed by only 2 to 3% for the three runs. The ranges of VOC removal efficiencies during the Phase 3 reproducibility runs are shown in Table 5.7.

A

EX-SITU ELECTROCHEMICAL TREATMENT PROCESSES

Ex-situ electrochemical treatment processes are intended to treat aqueous streams contaminated with metals, suspended solids, emulsions, and some organic compounds. These processes fall into two broad categories: electrochemical coagulation (electrocoagulation) and electrochemical oxidation/reduction. This appendix provides case studies for three ex-situ electrochemical treatment processes using Electrochemical Coagulation and Alternating-Current Electrocoagulation:

1. ACE Separator™ marketed by ElectroPure Systems Inc. — This technology was the subject of a testing program under the emerging technology portion of US EPA's Superfund Innovative Technology Evaluation (SITE) program (Barkely, Farrell, and Williams 1993).
2. Electrochemical Treatment System (no trade name) — Andco Environmental Processes, Inc. evaluated a pilot-scale electrochemical treatment system at the Milan Army Ammunition Plant (Laschinger 1992).
3. Electrochemical Oxidation/Reduction — Silver (II) Process marketed by AEA Technology (Oxfordshire, UK). This process is based on the electrochemical cell used for chlorine production. The technology is applicable only to the treatment of very low-volume waste streams. It is being marketed for the treatment of highly toxic materials only such as nuclear waste and chemical munitions (Batey 1995).

A.1 Electrochemical Coagulation

Chemical coagulation has been used for decades to destabilize colloidal suspensions and to effect precipitation of soluble metal species as well as other inorganic species from aqueous streams. Alum, lime, and/or polymers have been the chemical coagulants used. These processes, however, tend to generate large volumes of sludge with a high bound-water content that can be slow to filter and difficult to dewater. The treatment processes also tend to increase the total dissolved solids content of the effluent, making it unacceptable for reuse within industrial applications.

Electrocoagulation uses an electric field to achieve the same effect as chemical coagulation, but without some of the previously mentioned drawbacks. Rather than adding chemical agents to a wastewater, electrochemical oxidation/reduction runs a current between two electrodes. The electrodes can be made of aluminum or iron or aluminum or iron pellets can be placed between the electrodes. In either case, the current generates aluminum or iron ions which, along with the electric potential, result in the coagulation of the suspended solids. The current also causes dissolved metals to precipitate. In its simplest form, an electrocoagulation system passes the aqueous stream past energized electrodes that modify the surface potential of the particulate, causing it to agglomerate into large particles that settle or filter more readily.

The alternating current electrocoagulation (ACE) technology was originally developed in the early 1980s to break stable aqueous suspensions of clays and coal fines produced in the mining industry. Traditionally, these effluents were treated with conventional techniques that made use of organic polymers and inorganic salts to agglomerate and enhance the removal of the suspended materials. The ACE technology was developed to simplify effluent treatment, realize cost savings, and facilitate recovery of fine-grained coal.

ACE is based upon colloidal chemistry principles — principles using alternating electrical power and electrophoretic metal hydroxide coagulation. The basic mechanism for the technology is electroflocculation wherein small quantities (generally <30 mg/L) of aluminum hydroxide species are introduced into solution to facilitate flocculation. Electroflocculation causes an effect similar to that produced by the addition of chemical coagulants such as aluminum or ferric sulfate. These cationic salts destabilize colloidal suspensions by neutralizing negative charges associated with these particles at neutral or alkaline pH. This enables the particles to come together closely enough to agglomerate under the influence of van der Waals attractive forces. See Figure A.1 for the ACE basic process flow.

Although the electroflocculation mechanism resembles chemical coagulation in that cationic species are responsible for the neutralization of surface charges, the characteristics of the electrocoagulated floc differ dramatically from those of floc generated by chemical coagulation. An electrocoagulated floc tends to contain less bound water, is more shear resistant, and is more readily filterable.

Application of an AC electric field to the electrodes induces dissolution of the aluminum and formation of the polymeric hydroxide species. Charge neutralization and particle growth are initiated within the electrocoagulation cells and continue following discharge of the aqueous medium from the apparatus. In this way, product separation into solids, water, and oils may be achieved.

Figure A.1
Schematic of an ACE Separator™ Used in Alternating-Current Electrocoagulation

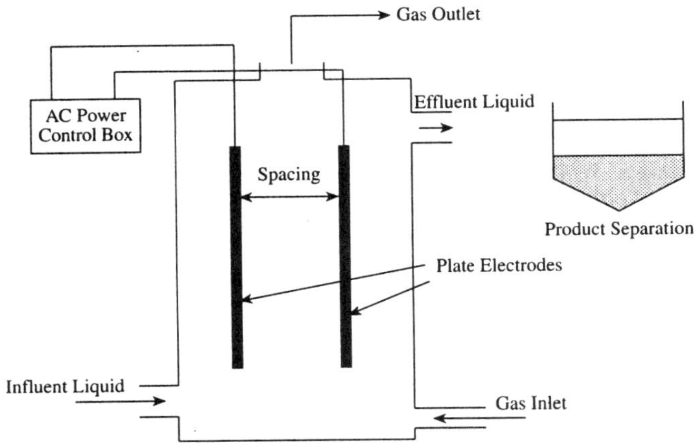

Source: Barkley, Farrell, and Williams 1993

A.1.1 ACE Technology

A two-year research effort was conducted by ElectroPure Systems, Inc., to evaluate the technical and economic feasibility of ACE for remediation of aqueous waste streams at Superfund sites (Barkley, Farrell, and Williams 1993). The ACE technology introduces low concentrations of nontoxic aluminum hydroxide species into the aqueous media by the electrochemical dissolution of aluminum-containing electrodes or pellets. The aluminum species that are produced neutralize the electrolytic charges on suspended material and/or prompt the coprecipitation of certain soluble ionic species, facilitating their removal.

A.1.1.1 Process Description

The ACE technology was tested using two designs of the ACE Separator™:

- a parallel-electrode unit in which a series of vertically-oriented aluminum electrodes form a series of monopolar electrolytic cells through which the effluent passes; and
- a fluidized-bed unit with nonconductive cylinders equipped with nonconductive metal electrodes between which a turbulent fluidized bed of aluminum alloy pellets is maintained.

Electrocoagulation operating conditions are highly dependent on the chemistry of the aqueous medium, especially conductivity. Other characteristics, such as pH, particle size, and chemical constituent concentrations will influence operating conditions. Treatment generally requires application of low voltage (<150 VAC) to the electrocoagulation cell electrodes; current usage is typically 1 to 5 amp-min/L (4 to 19 amp-min/gal). The flow rate of the aqueous medium through an electrocoagulation cell depends on the electrical conductivity of the solution, the nature of the entrained suspension or emulsion, and the extent of electrocoagulation required to achieve the treatment objective. Retention times as short as 5 sec are sometimes sufficient to break a suspension. Electrocoagulation may be accomplished in a single pass or multiple passes (recycle mode).

In the fluidized-bed unit, compressed air is introduced into the electrocoagulation cells to assist in maintaining the turbulent fluidized bed and to enhance the aluminum dissolution efficiency by increasing the anodic surface area. It also provides a mechanical scrubbing action within the electrocoagulation cell that reduces buildup of impermeable oxide coatings on the aluminum pellets and the inherent loss of efficiency that would result. Typically, the fluidized-bed unit dissolves aluminum at least one order of magnitude more efficiently than the parallel-electrode unit. Depending on system configuration, maintenance of the apparatus is limited to periodic replenishment of the aluminum fluidized-bed material and/or electrodes. For most applications, pellets for the fluidized-bed unit can be produced from recycled aluminum scrap or beverage containers. Where sludge reclamation is the objective, however, the use of higher quality pellets is required to reduce the introduction of impurities in the sludge.

A.1.1.2 Technology Testing

ElectroPure Systems, Inc. tested the technology in both the parallel-electrode and fluidized-bed configurations on various surrogate wastes containing emulsified diesel fuel metals and clays. The wastes were prepared to resemble those from leaking from underground storage tanks and soil washing operations. The primary objective of such testing was to establish operating conditions for the ACE Separator™ to break the oil/water emulsion and achieve reductions in clay, suspended solids, and soluble metal pollutant loadings.

The surrogate wastes were prepared by mixing 0.2 to 3.0% (by weight) of the -230 mesh (clay and silt) fraction of the US EPA's synthetic soil matrix (SSM) with the following:

- 0.5 to 1.5% (by weight) Number 2 diesel fuel;
- 0.05 to 0.10% (by weight) of an emulsifier (Titon-100X* or Alconox soap); and
- 10 to 100 mg/L of one or more of the following contaminants: copper, nickel, zinc orthophosphate, or fluoride.

The pH of each surrogate mixture was adjusted with either sodium hydroxide or calcium oxide to the desired value (5, 7, or 9) and the conductivity was raised to roughly 1200 to 1500 μS/cm (3,000 to 3,800 μS/in.) with sodium chloride to simulate values expected in nature.

Initially, bench-scale electrocoagulation experiments using the parallel electrode unit were conducted on five aqueous-based systems that included a metals mixture, clay suspension, diesel fuel emulsion, soluble organic salt, and diesel fuel/soluble organic emulsion. Optimum treatment times were established by examining the contaminant loadings as a function of treatment time. To compare the results with conventional treatment processes, aliquots of each surrogate stock solution were treated with alum. Sufficient alum was added to give the aluminum equivalent to that introduced by the electrocoagulation equipment.

When the results of the bench-scale experiments were applied to flow reactor testing during the second year of the program, the following operating difficulties were encountered:

- persistent electrode coating and fouling; and
- low efficiencies of aluminum generation.

The program was, therefore, modified to include testing with the fluidized-bed electrocoagulation cell design. Three phases of laboratory experiments were undertaken to evaluate both electrocoagulation units: (1) preliminary screening experiments to demonstrate the feasibility of reducing the concentration of each metal, (2) matrix experiments to define the most opportune retention time and current (or current density), and (3) optimization experiments to define other ACE Separator™ operating parameters to achieve the most cost-effective removal conditions. The pH was adjusted to 5, 7, or 9 and the conductivity raised to approximately 1200 μS/cm (3,000 μS/in.)with sodium chloride. The conductivity of some surrogate wastes was increased to approximately 3000 μS/cm (7,600 μS/in.) and subjected to electrocoagulation. Surrogate wastes subjected to these experiments included the five aqueous systems listed above as well as surrogate wastes containing individual constituents such as nickel, zinc, copper, fluoride, and phosphate.

Optimum operating conditions for the parallel-electrode unit were developed from these studies (Table A.1). These conditions served as the basis for the subsequent pilot-scale tests. Similarly, the optimum operating conditions for the fluidized-bed unit were 2.54 cm (1 in.) electrode spacing, 8- to +16-mesh aluminum pellets size, and 20 amp current.

The bench-scale experiment conducted on the US EPA surrogate wastes led to the following findings:

- when compared with alum treatment, electrocoagulation formed approximately 83% less sludge volume and the filtration rate improved to 76%;
- for the fluidized-bed configuration, aluminum or stainless steel may be used as electrode material with comparable results; and

- with both increased frequency for the AC and increased retention time, the agglomerated particles tend to disaggregate.

Pilot-scale tests were performed using both the parallel and fluidized-bed configurations of the ACE Separator™. A 12 hour experiment using the ACE fluidized bed separator™ was conducted on 208 L (55 gal) batches of surrogate waste solution containing 0-2% (by weight) SSM fines, 0.5% (by weight) diesel fuel, 0.05% (by weight) Alconox surfactant, and 10 mg/L each of Cu^{2+}, Zn^{2+}, PO_4^{3-}, F^-, and Ni^{2+}. The conductivity and pH of the solution were raised to 1,200 µS/cm (3,000 µS/in.) and 7, respectively. The surrogate was recycled through a 10.2-cm (4-in.)-diameter, Schedule-80 PVC pipe, 61.0-cm (24-in.)-high pilot-scale ACE Separator™ that was equipped with two Type 316 stainless-steel electrodes 61-cm-high, 6.4-cm-wide (24-in.-high, 2.5-in.-wide) and whose interior was filled with 8-to +16-mesh aluminum pellets. The unit was powered at a constant 20 amp, and the voltage was allowed to vary as the electrocoagulation treatment progressed over the 12 hour period. In this experiment, the flow of the surrogate solution through the ACE Separator™ was varied from 3.8 to 22.7 L/min (1 to 6 gal/min) and the quantity of compressed air introduced into the solution feed line was a maximum of 10 psig (0.07 MPa). Samples of the surrogate solution were collected at various times throughout the experiment to document the rate of aluminum ion generation and the reductions in concentration of the metal contaminants, chemical oxygen demand (COD), and total suspended solids (TSS).

Table A.1

Optimum Operating Conditions for Parallel Electrode Unit Based on Bench-Scale Tests

Parameter	Value
Current	4 amp
Electrode Spacing	1.27 cm (0.5 in.)
Retention Time	3 to 5 min
Frequency	10 Hz
Submergence	Fully submerged

Source: Barkley, Farrell, and Williams 1993

In a similar pilot-scale test using the parallel-plate unit, the surrogate waste was composed of essentially the same constituents as for the fluidized-bed experiment. The notable changes were that the conductivity of the solution was increased to approximately 3,000 μS/cm (7,600 μS/in.)and no fluoride salt was added. The other operating parameters were based on results obtained from the bench-scale tests. The aluminum generation and consumption rates and the electrical power required to effect acceptable phase separation as well as contaminant reductions were monitored.

Throughout the various phases of the experimental program, samples of the treated effluent were collected and allowed to settle for 30 minutes. The supernate was removed and analyzed. The subnate, containing the settled floc, was filtered and the filtrate and filter cake were analyzed.

Pilot-scale tests were conducted on both the parallel and fluidized-bed configurations of the ACE Separator™ on a 3% soil slurry containing roughly 50% clays, 1.5% diesel fuel and 0.1% of a strong surfactant. Electrocoagulation reduced TSS from 22 mg/L to 4.5 mg/L and total organic carbon (TOC) from 130 mg/L to 6.6 mg/L. Copper was reduced by 72%, cadmium by 70%, chromium by 92%, and lead by 88%. No appreciable change in total solids (TS) loading in the supernate resulted from electrocoagulation.

Particle size was enhanced in the clay fraction as a result of electrocoagulation. The mean particle diameters of the ACE Separator™ treated particulate, both in the supernate and in the filtrate (188 μm and 230 μm [$7.4 \cdot 10^{-3}$ in. and $9.1 \cdot 10^{-3}$ in.], respectively) increased by a factor of approximately 85 and 105 respectively over that in the original slurry (2.2 μm [$0.9 \cdot 10^{-3}$ in.]).

Data obtained from the 12 hour, pilot-scale, fluidized-bed test revealed that after 30 minutes of treatment, more than 90% of the metals and phosphates were removed. Aluminum generation rates were highest when the throughput flow rate was less than 15 L/min (4 gal/min). This upper flow limit may reflect compaction of the fluidized bed aluminum pellets against the upper screen of the electrocoagulation cell, thus placing them out of the range of the electrodes. As the emulsion is destabilized, the surrogate solution most likely becomes less resistive to ion mobility and, thereby, improves the operational efficiency of the ACE Separator™.

Filtration time for solids coagulated from particulate suspensions and oily emulsions by electrocoagulation is much less than for solids coagulated by chemical addition. Slurries tested were treated with alum addition and with an ACE Separator™. Electrocoagulation improved the filtration rate of titanium oxide by 63%. Other examples (for an oily emulsion and for biological sludge) indicate highly enhanced filtration rates for electrocoagulated wastewaters. Shear strength of an electrocoagulation floc is generally much greater than the shear strength of an alum floc. Both sonic treatments (used to evaluate the structural integrity of the floc) and actual filtration tests demonstrated high shear strength of the electrocoagulated flocs.

Electrocoagulation of metal- and phosphate-bearing industrial solutions indicates excellent nickel, copper, and phosphate reductions. More than 90% (concentration basis) of phosphate and copper can be removed from such solutions at low aluminum and electric power requirements. Reduction in the nickel concentration varies between 75% and 55% (concentration basis).

Electrocoagulation of synthetic laboratory solutions and industrial wastewater also confirmed the feasibility of using electrocoagulation for phosphate removal. Treatment of effluent from a commercial laundry reduced the phosphate concentration (PO_4^{2-}) from 45 mg/L to 5.4 mg/L after low-intensity electrocoagulation (0.36 kW, 0.75 min retention time). Electrocoagulation of process water from a phosphate mining operation reduced the phosphate level by 91%, from 160 mg/L to 14 mg/L (3.3 kW, 0.17 min retention time). Finally, treatment of dilute phosphoric acid solutions with a nominal 100 mg/L total phosphate concentration and a conductivity of approximately 2000 μS/cm resulted in greater than 95% reductions in soluble phosphate over a range of acidities.

A.1.1.3 Costs

Based on bench-and pilot-scale testing, projected treatment cost estimates were developed. Overall treatment operating costs (electricity, aluminum pellets, operation, and maintenance) will vary upwards from $0.13/1,000 L ($0.50/1,000 gal), depending on emulsion strength, unwanted component concentration(s)(e.g., emulsifiers) in the effluent, and effluent TSS. Additional cost considerations may be involved in full-scale operation. Operator supervision and maintenance would be limited to periodic replenishment of the aluminum pellets, chemical pretreatment systems (e.g., salt addition to enhance conductivity), and electrode replacement. Estimated operating costs are based on laboratory and limited pilot-scale testing of effluents; currently, these costs exceed those for comparable traditional chemical treatment (alum or polyelectrolytes). The lower maintenance and operator supervision required for ACE Separator™ operation and the capability to use ACE Separator™ treated water in closed-loop, zero-discharge applications adds to its attractiveness. Successful commercialization requires further research to significantly improve aluminum dissolution efficiency. If the ACE Separator™ can be engineered to regularly generate sufficiently high aluminum dissolution concentrations, the technology may be applicable to industrial effluent treatment trains, as well as to some site remediation activities.

The capital cost for a standard ACE Separator™ with a nominal throughput capacity of 190 L/min (50 gal/min) is estimated by the vendor at $80,000; for a 946 L/min (250 gal/min) throughput, $300,000.

A.1.1.4 Conclusions

The technology offers a promising alternative for treating waste streams containing clays, certain metal constituents, and other soluble pollutants. As an alternative to chemical conditioning, the technology appears to have an advantage over chemical coagulation because it does not add extraneous soluble solids and because the sludge has a lower water content. As a result, the sludge from an electrocoagulation process has better filtering characteristics than a sludge from a chemical coagulation process. The effectiveness of electrocoagulation as compared to alum addition and polymer coagulation offers:

- TSS: Electrocoagulation treatment and the polymer treatment yielded equivalent results for the reduction of TSS in the treated supernate. TSS values for alum treatment were four to five times greater than those for ACE Separator™ treatment or polymer treatment.
- Chemical Oxygen Demand (COD): Electrocoagulation resulted in the highest COD reductions of the three methods. Removal efficiency for COD was from two to four times higher than removal efficiency for either alum treatment or polymer treatment.
- Lead: At high concentrations of lead, electrocoagulation achieved approximately 55% removal of lead whereas, polymer treatment showed higher removal (71%). Because some difficulties were experienced with the alum treatment, these test results were invalidated. Electrochemical treatment of slurries with low concentrations of lead resulted in the highest removal (96%).

- Copper: The removal efficiency for copper in the supernate was very high using electrocoagulation; 90% reduction was observed when the concentration was high and 99% reduction was observed when the concentration of copper was low. For high concentrations of copper, however, polymer and alum addition achieved greater removal — virtually 100% reduction.
- Chromium: Electrochemical treatment resulted in good removal for total chromium (87% and 94% reduction for the high and low concentrations, respectively). Alum and polymer addition accomplished similar removal.
- Cadmium: Cadmium levels in the supernate dropped as a result of electrochemical treatment —14% in when the concentration was high and 99% in when the concentration was low. The inconsistency between these two sets of experiments, as well as the high concentrations remaining in the supernate and filtrates, raised questions about the accuracy of the results for the high concentration tests. For the low concentration tests, the cadmium concentrations in both filtrates were much lower than the concentrations for either the alum or polymer-treated waters.

The following generalizations on the effectiveness of the ACE treatment are made:

- ACE Separator™ treatment consistently reduces the TS and TSS loadings to a degree equivalent with polymer treatment and to approximately one-quarter the level achieved through alum addition; and
- better reductions in soluble metal concentrations are achieved with electrocoagulation treatment than with alum treatment.

In summary, electrocoagulation is a promising, technically simple method for achieving solid-liquid separations in aqueous-based waste streams. The majority of the nontoxic, aluminum ionic species introduced will be removed in the coagulated solids phase. The technology may be particularly suitable for zero-discharge applications where the addition of chemicals and the buildup of residual dissolved solids would adversely affect effluent quality or inhibit effluent reuse. Other potential applications include: (1) remediation of groundwater and leachates (metals, COD/BOD removal), (2) enhancement of clay separation from aqueous suspensions or emulsions resulting from soil washing operations, (3) breakage of oil/water emulsions produced in the pumping of hydrocarbon-contaminated groundwater, (4) removal of TSS from stormwater runoff, and (5) separation of oils and contaminants from thermal treatment system condensate.

A.1.2 Andco's Electrocoagulation Pilot Study

Andco Environmental Processes undertook a pilot-scale study of electrocoagulation to remove heavy metals and suspended solids from the Milan Army Ammunition Plant (MAAP) located in the City of Milan about 32 km (20 miles) north of Jackson, Tennessee. The MAAP has packaged and manufactured ordnance for over 50 years. At the time of operation, it was acceptable practice to rinse the packaging facilities with a deluge of water to keep the facilities free of spilled explosives. The rinse water, with high concentrations of explosives, was sent to lined ponds. The ponds have since been capped and a carbon adsorption treatment system is now being used to treat any packaging plant process water. This pilot study was initiated to find a suitable technology for cleaning up the large volume of groundwater that became contaminated as a result of the pond water leaching into the ground.

The water to be treated was contaminated with both metals and organics at concentrations above standard discharge limits. The pilot study consisted of two independent processes. A pretreatment step for removal of metals and any suspended solids by electrocoagulation and a secondary UV ozone process for oxidation of organics. This discussion focuses on the electrocoagulation system that was supplied by Andco to remove heavy metals, primarily manganese and mercury, and suspended solids.

A.1.2.1 Process Description

The Andco process electrochemically generates iron hydroxide from steel electrodes. Through coprecipitation and adsorption, the iron hydroxide acts to remove manganese, mercury, and other heavy metals from solution by forming an iron hydroxide/heavy metal matrix. Electrochemical treatment is followed by pH adjustment, clarification, and filtration.

The electrochemical cell used in the pilot study consists of a fiberglass body which supports and maintains a small gap between the sacrificial steel electrodes. The system consists of two cells, each with 16 electrodes. A direct electrical current is applied to the end electrodes and passed from electrode to electrode throughout the process water flowing through the cell. The electrical current causes water to break down into hydrogen gas and hydroxyl ions with the simultaneous generation of ferrous ion from the steel electrodes. The net effect is the formation of ferrous hydroxide. Hydrogen gas is generated as a byproduct of the reaction and is released through a vent on top of the cells. The process water is then passed through a retention tank to remove the remaining entrained hydrogen bubbles.

From the retention tank the process water flows to the pH adjustment tank where a sodium hydroxide solution is added to increase the pH to between 9.0 and 9.3. The pH of incoming groundwater during the tests was typically 5.8. An increase in the process water pH occurs in the electrochemical cell. The ferrous hydroxide generated in the cells is a weak base. Upon exiting the cells, the pH of the test water was approximately 6.

Following the pH adjustment tank, the water flows by gravity to a corrugated, inclined-plate clarifier. A small amount of polymer flocculent is mixed with the process water in the flash mix chamber to improve the settling characteristics of the precipitated solids. Next, the water flows to the flocculator, where a picket fence-type mixer gently agitates the solids, causing collisions that form larger solids. The solids settle to the bottom of the clarifier, and the clear water overflows the effluent weir. Due to the low solid content of the process water entering the clarifier, a sludge recycle pump is included. The sludge recycle pump draws settled sludge from the cone of the clarifier and pumps it to the flash mix chamber. This provides the clarifier with a higher solids content to improve floc quality and clarifier performance.

The settled solids are occasionally pumped from the cone of the clarifier to the sludge holding tank before being sent to the filter press. The filter press takes the 1-2% solids sludge from the sludge holding tank and produces a 25-30% solids filter press cake. The overflow from the clarifier flows to a surge tank and is pumped through a polishing, multi media filtration system. The process water is sent to a treated water holding tank. Here, the pH is adjusted to neutral before being sent to the UV ozone system, or discharged to the treated water holding pool.

A.1.2.2 Treatment Levels

The treatment levels chosen for the pilot study were based on the result of bench-scale treatment tests. For days 3 through 5, the treatment goal was 25 mg/L of iron. On day 6, a higher treatment goal of 50 mg/L was used. Samples were taken regularly just after the electrochemical cell and the iron level was determined using a HACH DR2000 Spectrophotometer. Table A.2 contains the average treatment levels based on the daily average iron generation. The iron generation was compared to the theoretical generation based on the cell amperage and flow to determine the efficiency.

Table A.2
Electrochemical Iron Treatment Levels

Day	Treatment Level (ppm Fe)	E/C Cell Efficiency (%)
2	–	–
3	28	135
4	32	157
5	32	157
6	50	158

Source: Barkley, Farrell, and Williams 1993

On day 2, the iron generation was erratic and inconsistent; it was not possible to accurately determine the average iron generation for that day. In general, the iron generation was well below the theoretical level for the grab samples collected on day 2. It appeared that the process water chemistry was passivating the electrodes and reducing the cell's efficiency. On days 3 through 6, a salt (sodium chloride) was added to change the process water chemistry. A small amount of salt greatly increased cell performance.

Table A.3

Electrochemical Precipitation — Days 2 and 3 Pilot-Scale Treatablility Data

Parameter Analyzed	DAY 2		DAY 3				Treatment Goals	
	Untreated Water	Electrochemical Precipitation	Untreated Water		Electrochemical Precipitation		Reinjection (µg/L)	Surface Water Discharge (µg/L)
	PRETRT2 (µg/L)	PRECIP2 (µg/L)	PRETRT3 (µg/L)	PRETRT3A (µg/L)	PRECIP3 (µg/L)	PRECIP3A (µg/L)		
TAL Metals								
Aluminum	<141	<141	<141	<141	<141	<141	NTG	NTG
Antimony	<38.0	<38.0	<38.0	<38.0	<38.0	<38.0	–	–
Arsenic	<2.54	<2.54	<2.54	<2.54	<2.54	<2.54	0.0175	0.0175
Barium	89.7	55.5	95.3	91.1	30.9	26.1	1,000	1,000
Beryllium	<5.0	<5.0	<5.0	<5.0	<5.0	<5.0	–	–
Cadmium	<4.01	<4.01	<4.01	<4.01	<4.01	<4.01	–	–
Calcium	16,300	16,100	16,500	16,300	15,500	15,200	NTG	NTG
Chromium	<6.02	<6.02	<6.02	<6.02	<6.02	<6.02	50	11
Cobalt	<25.0	<25.0	<25.0	<25.0	<25.0	<25.0	–	–
Copper	8.60	<8.09	<8.09	<8.09	<8.09	<8.09	1,300	6.54
Cyanide	13.4	10.1	9.54	13.7	4.31	18.6	200	5.2
Iron	<38.8	2,570	<38.8	<38.8	214	254	NTG	300
Lead	1.8	1.7	1.4	<1.3	<1.3	<1.3	15	1.32
Magnesium	5,700	5,630	5,750	5,820	5,410	5,330	NTG	NTG
Manganese	820	196	838	845	15.1	14.5	NTG	50
Mercury	0.5	<0.2	0.6	0.5	<0.2	<0.2	1.1	0.012
Nickel	<34.3	<34.3	<34.3	<34.3	<34.3	<34.3	100	88
Potassium	1,310	1,430	1,490	1,600	1,480	1,550	NTG	NTG
Selenium	<3.0	<3.0	<3.0	<3.0	<3.0	<3.0	–	–
Silver	<4.60	<4.60	<4.60	<4.60	<4.60	<4.60	50	1.2

Table A.3 cont.
Electrochemical Precipitation — Days 2 and 3 Pilot-Scale Treatability Data

Parameter Analyzed	DAY 2		DAY 3				Treatment Goals	
	Untreated Water	Electrochemical Precipitation	Untreated Water		Electrochemical Precipitation		Reinjection (µg/L)	Surface Water Discharge (µg/L)
	PRETRT2 (µg/L)	PRECIP2 (µg/L)	PRETRT3 (µg/L)	PRETRT3A (µg/L)	PRECIP3 (µg/L)	PRECIP3A (µg/L)		
TAL Metals								
Sodium	6,470	19,500	7,100	6,730	46,200	46,200	NTG	NTG
Thallium	<81.4	<81.4	<81.4	<81.4	<81.4	<81.4	–	–
Vanadium	<11.0	<11.0	<11.0	<11.0	<11.0	<11.0	–	–
Zinc	47.1	<21.1	60.3	31.6	<21.1	<21.1	2,000	59
TCL Volatiles								
Acetone	NA	NA	<13	<13	NA	NA	NTG	NTG
Benzene	NA	NA	<0.50	<0.50	NA	NA	–	–
Bromodichloromethane	NA	NA	<0.59	<0.59	NA	NA	–	–
Bromoform	NA	NA	<2.6	<2.6	NA	NA	–	41
Bromomethane	NA	NA	<5.8	<5.8	NA	NA	–	–
(2-Butanone) Methyl Ethyl Ketone	NA	NA	<6.4	<6.4	NA	NA	–	–
Carbon Disulfide	NA	NA	<0.50	<0.50	NA	NA	NTG	NTG
Carbon Tetrachloride	NA	NA	<0.58	<0.58	NA	NA	–	–
Chlorobenzene	NA	NA	<0.50	<0.50	NA	NA	–	–
Chloroethane	NA	NA	<1.9	<1.9	NA	NA	–	–
2-Chloroethylvinyl Ether	NA	NA	<0.71	<0.71	NA	NA	–	–
Chloroform	NA	NA	<0.50	<0.50	NA	NA	100	0.19

Chloromethane	NA	NA	<3.2	<3.2	NA	NA	–	–
Dibromochloromethane	NA	NA	<0.67	<0.67	NA	NA	–	–
Dichlorobenzene	NA	NA	<10	<10	NA	NA	–	–
1,1-Dichloroethane	NA	NA	<0.68	<0.68	NA	NA	–	–
1,2-Dichloroethane	NA	NA	<0.50	<0.50	NA	NA	–	–
1,1-Dichloroethylene	NA	NA	<0.50	<0.50	NA	NA	–	–
1,2-Dichloroethene	NA	NA	<0.50	<0.50	NA	NA	–	–
1,2-Dichloropropane	NA	NA	<0.50	<0.50	NA	NA	–	–
Cis-1,3-Dichloropropene	NA	NA	<0.58	<0.58	NA	NA	–	
Trans-1,3-Dichloropropene	NA	NA	<0.70	<0.70	NA	NA	–	–
Ethylbenzene	NA	NA	<0.50	<0.50	NA	NA	680	680
2-Hexanone	NA	NA	<3.6	<3.6	NA	NA	–	–
Methylene Chloride	NA	NA	<2.3	<2.3	NA	NA	–	–
Methyl Isobutyl Ketone	NA	NA	<3.0	<3.0	NA	NA	–	–
Styrene	NA	NA	<0.50	<0.50	NA	NA	–	–
1,1,2,2-Tetrachloroethane	NA	NA	<0.51	<0.51	NA	NA	–	–
Tetrachloroethene	NA	NA	<1.6	<1.6	NA	NA	–	–
Toluene	NA	NA	1.96	1.45	NA	NA	–	–
1,1,1-Trichloroethane	NA	NA	<0.50	<0.50	NA	NA	1,000	1,000
1,1,2-Trichloroethane	NA	NA	<1.2	<1.2	NA	NA	–	–
Trichloroethene	NA	NA	<0.50	<0.50	NA	NA	–	–
Trichlorofluoromethane	NA	NA	<1.4	<1.4	NA	NA	–	–
Vinyl Acetate	NA	NA	<8.3	<8.3	NA	NA	–	–
Vinyl Chloride	NA	NA	<2.6	<2.6	NA	NA	–	–
Xylene	NA	NA	<0.84	<0.84	NA	NA	10,000	10,000
Acrolein	NA	NA	<100	<100	NA	NA	–	–
Acrylonitrile	NA	NA	<100	<100	NA	NA	–	–

– this chemical is not considered a contaminant of concern
NTG no treatment goal for this chemical
NA analyte not analyzed

Source: Laschinger 1992

Table A.4

Electrochemical Precipitation — Days 4, 5, and 6 Pilot-Scale Treatablility Data

Parameter Analyzed	DAY 4		DAY 5		DAY 6		Treatment Goals	
	Untreated Water	Electrochemical Precipitation	Untreated Water	Electrochemical Precipitation	Untreated Water	Electrochemical Precipitation	Reinjection (µg/L)	Surface Water Discharge (µg/L)
	PRETRT4 (µg/L)	PRECIP4 (µg/L)	PRETRT5 (µg/L)	PRECIP5 (µg/L)	PRETRT6 (µg/L)	PRECIP6 (µg/L)		
TAL Metals								
Aluminum	<141	<141	<141	<141	<141	<141	NTG	NTG
Antimony	<38.0	<38.0	<38.0	<38.0	<38.0	<38.0	-	-
Arsenic	<2.54	<2.54	<2.54	<2.54	<2.54	EX06/24	0.0175	0.0175
Barium	90.9	30.4	93.8	31.6	87.8	23.6	1,000	1,000
Beryllium	<5.0	<5.0	<5.0	<5.0	<5.0	<5.0	-	-
Cadmium	<4.01	<4.01	<4.01	<4.01	4.60	<4.01	-	-
Calcium	16,300	15,400	16,700	15,500	15,900	14,600	NTG	NTG
Chromium	<6.02	<6.02	<6.02	<6.02	<6.02	<6.02	50	11
Cobalt	<25.0	<25.0	<25.0	<25.0	<25.0	<25.0	-	-
Copper	<8.60	<8.09	<8.09	<8.09	<8.09	<8.09	1,300	6.54
Cyanide	17.2	20.5	8.99	21.4	12.8	17.2	200	5.2
Iron	<38.8	203	<38.8	191	<38.8	232	NTG	300
Lead	1.7	<1.3	<1.3	3.8	<1.3	EX06/24	15	1.32
Magnesium	5,750	5,330	5,870	5,420	5,520	4,020	NTG	NTG
Manganese	832	23.1	863	15.2	794	13.7	NTG	50
Mercury	0.4	<0.2	0.5	<0.2	0.5	<0.2	1.1	0.012
Nickel	<34.3	<34.3	<34.3	<34.3	<34.3	<34.3	100	88
Potassium	1,820	1,620	1,600	1,840	1,740	1,020	NTG	NTG

Selenium	<3.0	<3.0	<3.0	<3.0	<3.0	EX06/24	-	-
Silver	<4.60	<4.60	<4.60	<4.60	<4.60	<4.60	50	1.2
Sodium	6,590	56,200	6,930	57,000	6,260	43,200	NTG	NTG
Thallium	<81.4	<81.4	<81.4	<81.4	<81.4	<81.4	-	-
Vanadium	<11.0	<11.0	<11.0	<11.0	<11.0	<11.0	-	-
Zinc	40.8	<21.1	43.2	<21.1	48.8	<21.1	2,000	59
Explosives								
1,3-Dinitrobenzene	NA	NA	62.5	NA	NA	NA	1	1
2,4-Dinitrotoluene	NA	NA	136	NA	NA	NA	0.5	0.5
2,6-Dinitrotoluene	NA	NA	<0.738	NA	NA	NA	0.0068	0.0068
HMX	NA	NA	744	NA	NA	NA	400	400
Nitrobenzene	NA	NA	42.4	NA	NA	NA	17.5	17.5
RDX	NA	NA	4,110	NA	NA	NA	2	2
Tetryl	NA	NA	<15.6	NA	NA	NA	NTG	NTG
1,3,5-Trinitrobenzene	NA	NA	1,340	NA	NA	NA	2	2
2,4,6-Trinitrotoluene	NA	NA	11,000	NA	NA	NA	2	2
Other Parameters								
Nitrate/Nitrite	NA	NA	NA	NA	21,000	NA	10,000/1,000	NTG
Carbon								
Total Organic	NA	NA	NA	NA	12,400	NA		
Total Inorganic	NA	NA	NA	NA	2,400	NA	?	?
Purgeable Organic	NA	NA	NA	NA	<300	NA		

– this chemical is not considered a contaminant of concern
NTG no treatment goal for this chemical
NA analyte not analyzed

Source: Laschinger 1992

For days 3 through 6, the iron generation efficiency was well above 100%. In systems with low pH process water, iron generation efficiencies greater than 100% are sometimes observed. The added iron is due to the process water corroding the steel electrodes and, hence, dissolving additional iron that is added by the electric current in the cell. The observed iron generation rates were still considered unusually high, even considering the possibility of corrosion. It is doubtful that the corrosion effect alone would have such a significant effect on efficiency; other factors in the process water probably contributed.

A.1.2.3 Test Results

The treatment results for the electrochemical treatment are shown in Tables A.3 and A.4. Results were collected for days 2 through 6 (the first day was spent setting up the system). The heavy metal of primary concern, manganese, was to be reduced from an initial concentration of approximately 850 µg/L to less than 50 µg/L. The results for days 3 through 6 were very good, the residual manganese ranged from 13.7 to 23.1 µg/L. Only on day 2, was the result, 196 µg/L above the treatment goal. On this day, the residual iron was 2570 µg/L, which was also above the treatment goal of 300 µg/L. Day 2 was the first day of the pilot system operation, and iron generation difficulties were encountered. The overall treatment level was insufficient to form a good floc, and as a result, a portion of the iron stayed in a colloidal form and passed through the clarifier and multimedia filter. Also on the first day of operation, insufficient solids had built up in the clarifier to allow for sludge recirculation and increased floc quality. Precipitated manganese was probably carried along with the iron through the system. For all treated water tests, mercury was below detection limits of 0.2 µg/L.

An unusually high lead level, 3.8 µg/L, could have been due to an unusually high lead concentration in the influent, 1.8 µg/L. It could also have been caused by a short upset in the system operating pH that allowed the lead adsorbed on the iron floc to be released into the process water and show up as a spike in the effluent. However, if this were the case, other heavy metals would be expected to follow the same trend and this was not observed. The result may also have been due to inaccurate analysis.

Another unexpected trend in the results was that, in general, the cyanide level increased in the treated process water. No reaction occurred in the electrochemical cell that would produce cyanide. However, the cyanide level was not a concern of the electrochemical treatment because it was to be destroyed in the subsequent oxidation process.

From days 3 through 6, the residual iron level ranged from 191 to 254 μg/L. Although this was below the treatment goal of 300 μg/L, it was still more than expected and more than Andco typically encountered for ferric precipitation with polishing filtration. The expected residual iron concentrations is in the range of 50 μg/L or less.

During the pilot study, visible particles were regularly seen in the effluent of the multimedia filter The particles were large enough to be easily visible and should have been removed by the polishing filter. On two occasions, the filter was inspected and appeared normal. It is believed that a portion of the process water was bypassing the filtration bed due to the design of the multimedia filter valve body. Better performance will be achieved with a full-scale filter due to the use of discrete flow control valves and a deeper media bed.

A.1.2.4 Chemical Consumption

The pilot study used steel electrodes, sodium hydroxide, polymer flocculent, hydrochloric acid, sodium chloride, and hydrogen peroxide. The flow rate through the electrochemical cells was determined using an industrial rotameter with a stated accuracy of 15% of full-scale (Omega Model FL75).

A total of 67,090 L (17,725 gal) of contaminated groundwater was processed by the treatment system. Although the pilot system was initially operated at a 57 L/min (15 gal/min) feed rate, after two days of operation, the feed rate was cut to 51 L/min (13.5 gal/min) to match the maximum capacity of the pump supplying the contaminated water to the tank. Matching the treatment flow rate with the feed rate enabled for continuous operation of the pilot system.

The chemical consumption for a full-scale system was calculated based on the operation parameters of the pilotstudy. Table A.5 contains these values calculated on a per-million gallons (3,785,344 L) treated basis.

The solids generated and precipitated by the electrochemical process were removed in the form of a filter press cake. The filter press cake was composed mostly of ferric hydroxide and also included manganese, mercury, other heavy metals, suspended solids, and other components adsorbed by the ferric floc. While no leaching tests were performed on the sludge, the vendor contends it would pass the TCLP, and be classified as a nonhazardous waste. At a treatment level of 25 mg/L of electrochemically-generated iron and assuming that 10 mg/L of other components are removed from the process water, 0.6 m^3 (22 ft^3) of sludge will be produced per million gallons treated. This value assumes the filter can press cake to be 30% solids and have a density of 1,281 kg/m^3 (80 lb/ft^3).

Table A.5
Chemical and Electrical Power Consumption
Per-Million Gallons Treated, AndcoSystems

Item	Usage	Unit Cost ($)	Total Cost ($)
Steel Electrodes at 28 mg/L Fe	291 lb	0.039/lb	11.35
Cell Power at 28 mg/L Fe	326 kWh	0.065/kWh	21.19
Sodium Hydroxide	385 lb	0.065/lb @ 50%	25.03
Polymer Flocculent	10 lb	1.45/lb emulsion	14.50
Hydrochloric Acid	58 lb	0.12/lb @ 31%	6.96
Hydrogen Peroxide	20 lb	0.236/lb @ 50%	4.72
Sodium Chloride	188 lb	0.034/lb	6.39
Power Consumption for Pumping and Controls	1150 kWh	0.065/kWh	74.75
Total			164.89

Source: Laschinger 1992

The treatment level on day 3, 28 mg/L Fe, was the minimum treatment level used during the pilot-study, and at this level, the pilot system showed good results. It is probable a full-scale system operated at somewhat less than 25 mg/L Fe would meet treatment goals.

The electrodes used in the electrochemical process consisted of a commercial-grade cold rolled steel. The estimated electrode consumption assumes that the electrodes would be 80% consumed before replacement. The cell power represents the electrical power consumed by the DC power supplies in putting the 25 mg/L of iron in solution.

Sodium hydroxide was used for pH adjustment and for neutralization of the acid wash solution. In a full-scale system, nearly all of the sodium hydroxide consumed would be used for pH adjustment. Only an estimated 11% would be used for acid wash solution neutralization. The groundwater being treated in the tests had a pH of approximately 5.8. The iron added in the electrochemical cells caused the pH to increase to approximately 6.5. In the pH adjustment tank, sodium hydroxide was added to increase the pH to a range of 9.2 to 9.5. Increasing the pH increased the efficiency of manganese removal. The sodium hydroxide used in the pilot system was pre-diluted to 6.9% to allow for better pH control. In a full-scale system, sodium hydroxide would be added from a 50% solution. The sodium hydroxide consumption is based on the total amount consumed for pH adjustment during the entire pilot study divided by the total gallons treated.

The polymer flocculent used was Andco 2600 high-molecular weight, high-charge density emulsion. The polymer was diluted from its emulsion form to a 0.2% solution for metering into the process. The volume of 0.2% solution prepared and the volume remaining after the pilot study were recorded to determine the total consumption. The overall polymer addition was 1.24 mg/L.

Hydrochloric acid was used to make up the acid wash solution. The acid wash was used to recirculate a dilute acid solution through the electrochemical cells to remove any sludge and scale buildup from the electrodes. Excessive buildup in a cell can adversely affect the electrochemical reaction. At the end of the acid wash sequence, the acid solution was recovered for use in subsequent washes. In a full-scale system, the acid wash cycle is automated, and the acid solution strength is checked regularly and replenished on a monthly basis. Spent acid solution is neutralized and treated by the system. The pilot system acid wash solution was prepared by mixing 95 L (25 gal) of water and 15 L (4 gal) of 31% hydrochloric acid. A daily acid wash was performed at the start of each day. The acid consumption shown in Table A.4 describes the predicted consumption for a full-scale system operating at 1,900 L/min (500 gal/min). It is based on preparing a monthly 1,100 L (300 gal) batch of 10% hydrochloric acid plus an additional 50% to account for acid that would be added during the month to maintain the solution's strength.

A small amount of hydrogen peroxide was metered into the pH adjustment tank to increase manganese removal efficiency. Hydrogen peroxide is a strong oxidizer; it reacts with soluble manganese ions to form insoluble manganese dioxide. In initial lab tests, 10 mg/L of hydrogen peroxide was added to samples to ensure that all the iron generated by the electrochemical cell and the manganese would be completely oxidized. During the pilot study, the peroxide addition was originally set for 10 mg/L. This resulted in excessive residual peroxide which decomposed in the clarifier and caused a portion of the sludge to float. Therefore, the addition rate was decreased, resulting in a peroxide residual of 0.5 mg/L as measured in the pH adjustment tank The residual peroxide was regularly checked using EMQuant Peroxide Test Strips.

It was originally expected that peroxide would be consumed by the iron as it is oxidized from its +2 to +3 state. The iron added by the electrochemical cell was +2 or ferrous state. The pilot study revealed that there was enough dissolved oxygen and oxidizers in the process water to oxidize the iron to its ferric state before it entered the pH adjustment tank. This resulted in a lower-than-expected consumption of peroxide. The peroxide addition pump was calibrated to produce a 0.5 mg/L residual in the pH adjustment tank and remained at this setting until the final day of the pilot study when the dosing was increased to produce 2 mg/L peroxide residual. Based on the results, 0.5 mg/L residual peroxide achieved the desired treatment goals. The peroxide addition rate was determined to be 1.24 mg/L hydrogen peroxide based on a pumping rate calibration.

Sodium chloride was used to change the conductivity of the process water and, thereby, increase the iron generation efficiency. During the first day of system operation, the iron generation was erratic and inconsistent. Indications were that the electrodes were being passivated by the components in the process water. This situation is almost never encountered — the current applied to the electrochemical cells provides a strong driving force for the electrochemical reaction. Only under rare conditions has Andco encountered process water that can overcome the electrical driving force. The addition of a small amount of sodium chloride affects the conductivity of the water and the surface reaction at the electrodes. Sodium chloride is convenient to use because of its low hazard and low cost. In days 3,4, and 5 of the pilot tests, sodium chloride was added to the process water at a rate of 56 mg/L. On the last day of the pilot study, the salt addition was decreased to 23 mg/L. The addition of the salt greatly improved performance of the electrochemical cells. The performance did not decrease when the addition rate was decreased from 56 mg/L to 23 mg/L. The chemical cost information is based on the 23 mg/L addition. In a full-scale system, it is feasible that the minimum amount required is below 23 mg/L.

The power consumption estimate is based on the electrical power required for the process feed pumps, multimedia filter feed pumps, the pH adjustment tank, clarifier mixers, and system controls. The process feed pumps and the multimedia feed pumps consume approximately 85% of the electrical power. The power consumption for the process feed pump is based on a design pump discharge pressure of 2 atm (30 psi). The multimedia filter pump is based on a design discharge pressure of 2.4 atm (35 psi). The power consumption for an effluent pump was not included in this estimate because a full-scale system would most likely use a gravity-flow discharge. In determining the power consumption by the pumps, a motor efficiency of 80% and a pump efficiency of 60% were used.

A.2 Electrochemical Oxidation — Silver (II) Process

The Dounreay Silver (II) Electrochemical Oxidation Process for the destruction of organic wastes arose as a result of studies on the dissolution of intractable plutonium oxide residues created by the dissolution of nuclear (U, Pu) oxide fuel in nitric acid (Batey 1995). These intractable plutonium oxide residues could be taken into acid solution for eventual plutonium recovery, but to do so necessitated the use of particularly aggressive acid mixtures.

Experiments were performed using a simple, divided electrochemical cell where a solution of silver nitrate and nitric acid was placed in the anode compartment and nitric acid was placed in the cathode compartment. These experiments demonstrated that, with the passage of an electric current, intractable plutonium oxide residues dissolved rapidly. The Ag^{2+} ions generated at the anode were able to quickly oxidize the solid plutonium oxide to soluble PuO_2^{2+} and, at the same time, these ions were reduced to Ag^+ ions. The Ag ions could then be re-oxidized at the anode to Ag $(II)^+$ which could then react with insoluble material. The silver ions appeared to act as electron-transfer agents between the electric power being fed to the cell and the insoluble plutonium oxide, but were not consumed. This continuous use of the silver oxidant has permitted the development of a practical process that only requires the presence of a small amount of silver.

Based on the experiments, it was suggested that the Silver (II) would probably react with organic matter contaminated with plutonium, such as cellulose tissues used to clean up spills. Trials were carried out in which plutonium-contaminated tissues were placed in the anode compartment of an operating electrochemical dissolution cell. There was an immediate reaction as demonstrated by the disappearance of the dark brown Ag^{2+} ions, resulting in a clear solution. The process continued until all the tissues were consumed, whereupon the brown color of the Ag^{2+} ions again appeared. The cellulose tissues were completely oxidized to carbon dioxide and water.

It was then a relatively simple step to examine the possibility of using the process to destroy radioactive waste contaminated with tributylphosphate/odorless kerosene solvent from nuclear fuel reprocessing plants. The initial stages of these experiments in which solvent was added to the stirred compartment of an electrochemical cell were not encouraging. However, as the temperature in the cell increased due to the passage of the electric current, a reaction between the Ag^{2+} ions and the solvent was observed. At 55°C (131°F), the reaction with the electrochemically-generated Ag^{2+} ions resulted in the destruction of both the tributylphosphate and the kerosene. Oxidation of kerosene was surprising because it usually does not react with oxidizing agents.

The electrochemical cell used to produce Ag^{2+} ions is of the two-compartment type, with a fluoropolymer cationic-exchange membrane separating the anolyte and catholyte sections. The membrane is necessary because the reduced chemical species formed at the cathode, principally nitrous acid, would otherwise react with the silver (II) ions produced at the anode and reduce the efficiency of the destructive process. The anolyte is stirred or circulated to ensure that silver (I) ions are brought efficiently to the anode surface for oxidation to silver (II) ions; this transport process is the rate-limiting step. The silver (II) ions then either react directly with the organic material, or more likely, react with the water in the anolyte to form radical species such as •OH, which then in turn react with the organic material. The silver (II) ions are reduced to silver (I) ions in parallel with this reaction and must be oxidized at the anode for the destruction process to proceed to completion. In the case of the tributylphosphate/odorless kerosene solvent destruction, the final reaction products in the anolyte compartment are carbon dioxide, phosphate ions, and protons (i.e., water is consumed in the anolyte).

At the cathode, the nitric acid is reduced to nitrous acid (HNO_2), NO_x, and water; the precise chemistry is determined by the choice of electrode material. The formation of nitrous acid is the preferred reaction route as any further reaction results in gassing due to NO_x formation and may cause operational difficulties. The nitrous acid generated at the cathode can be converted back into nitric acid and recycled by a regenerative catholyte circulation system included in the process.

Two cell types that are manufactured by ICI, the filter press design and internally manifolded design, have been used to carry out the bulk of the studies performed. Small-scale studies employed the FMO1, a 1/35th scale model of the commercial-scale FM21SP electrochemical electrolyser and used a 60 amp bench-scale rig. This rig was used for the majority of the toxic organic destruction studies because of the small organic inventory required for operation. Process-scale studies employed the FM2l SF cell in a 2000 amp pilot rig. This latter rig was used to demonstrate the destruction on long runs of 10 days for tributyl phosphate/odorless kerosene and (up to 6 days) for organic ion-exchange resins.

The chemistry of the Silver (II) Process is summarized as follows:

1. At the anode, the silver (I) ions are oxidized to silver (II) ions:

$$6Ag^{+} \rightarrow 6Ag^{2+} + 6e^{-}$$

2. In the anolyte solution, the silver (II) ions react with water to form oxidizing species (•OH, •HO_2, •NO_3) represented by (O):

$$6Ag^{2+} + 3H_2O \rightarrow 6Ag^{+} + 3[O] + 6H^{+}$$

3. The oxidizing species then react with the organics in the waste stream that is introduced into the anolyte, oxidizing them to carbon dioxide, carbon monoxide, and water:

$$\text{“}CH_2\text{”} + 3[O] \rightarrow CO_2 + H_2O$$

"CH_2" represents a generalized carbon unit in an organic molecule, or more generally:

$$\text{Organics} + [O] \rightarrow CO_2 + CO + H_2O + \text{Inorganic Compounds}$$

When nitrogen, phosphorus, sulphur, or chlorine are present in an organic compound, these heteroatoms are oxidized to the mineral acid ion (e.g., nitrate, phosphate, sulphate, or chloride ions).

4. The silver (I) ions are then returned to the anode for reoxidization to silver (II) ions to enable the reaction to continue.

5. The protons in the form of hydronium ions (H_3O^+) migrate across the porous membrane to the cathode compartment under the influence of the applied voltage. The protons are consumed in the cathode reaction along with the nitrate ions to form (mainly) nitrous acid:

   The catholyte solution containing the formed nitrous acid is regenerated by reaction with oxygen. Thus, the overall stoichiometry of the process is:

$$\text{Organics} + O_2 \rightarrow CO_2 + H_2O + (\text{Inorganic Compounds})$$

Construction and operation of the Silver (II) process is illustrated by the following steps which refer to the simplified schematic of the process, Figure A.2.

(a) Chemical agent and makeup/feed chemicals are added to the nitric acid/silver (I) nitrate solution which forms the anolyte circuit (2) of the electrochemical cell (4).

(b) The anolyte solution is circulated through the electrochemical cell (4) where silver ions are transformed into silver (II) ions. These silver (II) ions attack the organic chemical agent and convert the organic chemicals to carbon dioxide, oxygen, trace NO_x (nitrogen oxides), protons, sulphate ions, phosphate ions, nitrate ions, and silver chloride. In this reaction, the silver (II) ions are reduced to silver (I) ions, which are recycled through the electrochemical cell to continuously generate silver (II) ions. Silver (I) ions, protons, and water diffuse through a cation exchange membrane within the electrochemical cell (4) to enter the catholyte circuit (3). The electrochemical cell (4) is the heart of the process and is a type used extensively in the chloralkali industry worldwide.

Figure A.2
Simplified Schematic of the Silver (II) Process

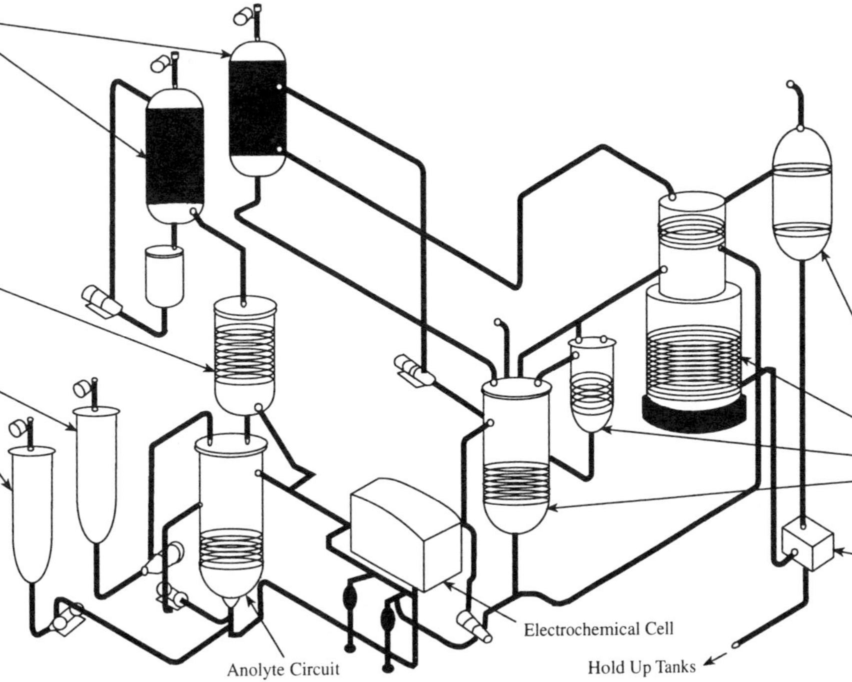

Reproduced courtesy of AEA Technology (Oxfordshire, UK)

(c) The catholyte circuit supports the balancing cathode reaction where nitric acid and protons are reduced to nitrous acid, NO_x, and water. The nitrous acid and NO_x are oxidized to nitric acid through reaction with oxygen and water. Excess water is removed by distillation and sampled to confirm the absence of chemical agents before discharge.

(d) Offgas from the anolyte circuit passes through a condenser (5) to remove water and nitric acid vapors. Condensate is returned to the anolyte circuit. The dried offgas stream is mixed with the offgas from the catholyte circuit and passed through a series of scrubbers (6) and an active charcoal filter to remove residual NO_x prior to discharge (6).

(e) At the end of a campaign, all of the solutions are discharged from the Silver (II) plant to a silver recovery plant (7). The final solutions are further tested for residual chemical agent prior to discharge.

While there appears little doubt that the Silver (II) process can totally oxidize organic compounds, the present systems are of limited size. The modules (as claimed by the vendor) use standard off-the-shelf chlorine industry cells and can be run in parallel to create a unit of any size desired. As such, the number of modules can be increased or upgraded as each application requires. Despite the drawbacks at present, this is one of the few processes that can oxidize organic compounds electrochemically at ambient pressures and low temperatures.

B

LIST OF REFERENCES

Acar, Y.B. 1992. Electrokinetic soil processing (a review of the state of the art). *Grouting, Soil Improvement and Geosynthetics, Geotechnical Special Publication.* 2(30): 1420-1423.

Acar, Y.B. 1993. Clean up with kilowatts. *Soils.* October: 38-41.

Acar, Y.B. and A.N. Alshawabkeh. 1993. Principles of electrokinetic remediation. *Environ. Sci. & Tech.* 27(13): 2638-2647.

Acar, Y.B., A.N. Alshawabkeh, and R.J. Gale. 1993. Fundamentals of extracting species from soils by electrokinetics. *Waste Management.* 13: 141-151.

Acar, Y.B., H. Li, and R.J. Gale. 1992. Phenol removal from kaolinite by electrokinetics. *J. Geotech. Eng.* 118(11): 1837-1852.

Acar, Y.B., J. Hamed, R.J. Gale, and G. Putnam. 1991. Acid/base distribution in electro-osmosis. *Transp. Res. Record.* (1288): 23-34.

Acar, Y.B., R.J. Gale, A.N. Alshawabkeh, R.E. Marks, S. Puppala, M. Bricka, and R. Parker. 1995. Electrokinetic remediation: basis and technology status. *J. Hazardous Materials.* 40: 117-137.

Allen, H.E. and P.H. Chen. 1993. Remediation of metal contaminated soil by EDTA incorporating electrochemical recovery of metal and EDTA. *Environmental Progress.* 12(4): 284-293.

Alshawabkeh, A. and Y.B. Acar. 1992. Removal of contaminants from soils by electrokinetics: a theoretical treatise. *J. Envir. Sci. and Health, Part A.* 27(7): 1835-1861.

American Public Health Association, American Water Works Association, and Water Environment Federation. 1992. *Standard Methods for the Examination of Water and Wastewater.* 18th edition. Washington, DC.

American Society of Civil Engineers. 1992. *Design of Municipal Wastewater Treatment Plants-Volume I and II.* WEF Manual of Practice No. 8, Water Environment Federation and ASCE Manual and Report on Engineering Practice No. 76.

American Society of Civil Engineers and American Water Works Association. 1990. *Water Treatment Plant Design.* 2nd edition.

Amirtharajah, A. and C.R. O'Melia. 1990. Coagulation processes: destabilization, mixing, and flocculation. *Water Quality and Treatment, A Handbook of Community Water Supplies.* American Water Works Association. 4th edition. F.W. Pontius (ed.). New York: McGraw-Hill.

Baillod, C.R., B.M. Faith, and O. Masi. 1982. Fate of specific pollutants during wet oxidation and ozonation. *Environmental Program.* 1(3): 217-227.

Barkely, N.P., C. Farrell, and T. Williams. 1993. *Emerging Technology Summary, Electro-Pure Alternating Current Electrocoagulation.* US EPA. USEPA/540/S-93/504. September.

Batey, B. 1995. The application of the electrochemical silver (II) process to the demilitarization of chemical munitions. Paper presented at AEA Technology Conference on Alternative Technologies for Chemical Weapon Demilitarization. Reston, VA. September 25-27.

Beeman, B. 1995. Personal communication with B. Beeman, Naval Communication Station, Stockton, CA. December.

Benefield, L.D. and J.S. Morgan. 1990. Chemical precipitation. *Water Quality and Treatment, A Handbook of Community Water Supplies*. American Water Works Association. 4th edition. F.W. Pontius (ed.). New York: McGraw-Hill.

Bettinger, J.A., E.D. Ferland, W. Killilea. 1994. Demonstration of the MODAR SCWO process. Paper Presented at Waste Management Meeting. Tucson Arizona. February 28-March 4.

Blaney, C.A., L. Li, E.F. Gloyna, and S.U. Hossain. 1995. Supercritical water oxidation of pulp and paper mill sludge as an alternative to incineration. *Innovations in Supercritical Fluids*. American Chemical Society. Symposium Series 608. K.W. Hutchenson and N.R. Foster (eds.). Washington DC.

Boltz, David F. and James A. Howell. 1979. *Colorimetric Determination of Nonmetals*. New York: John Wiley & Sons.

Bruell, C.J., B.A. Segall, and M.T. Walsh. 1992. Electro-osmotic removal of gasoline hydrocarbons and TCE from clay. *J. Env. Eng., ASCE*. 118(1): 68-83.

Buxton, G.V., C.L. Greenstock, W.P. Helman, and A.B. Ross. 1987. Critical review of rate constants for reactions of hydrated electrons, hydrogen atoms and hydroxyl radicals (•OH/•O) in aqueous solution. *Journal of Physical and Chemical Reference Data*. 17: 313-886.

Cabrera-Guzman, D., J.T. Swartzbaugh, and A.W. Weisman. 1990. The use of electrokinetics for hazardous waste site remediation. *J. Air Waste Manag. Assoc.* 40(12): 1670-1676.

Chen, C.T. 1995. Iron enhanced abiotic degradation of chlorinated hydrocarbons. Proceedings of the 21st Annual RREL Hazardous Waste Research Symposium. USEPA/600/R-95/012. April.

Cipollone, M.G., S.M. Hassan, N.L. Wolfe, D.R. Burris, and P.M. Jeffers. 1997. Reduction of halogenated hydrocrabons with iron: I. Kinetic observations. EPA Draft Report. Athens, GA.

Clifford, D.A. 1990. Ion exchange and inorganic adsorption. *Water Quality and Treatment, A Handbook of Community Water Supplies*. American Water Works Association. 4th edition. F.W. Pontius (ed.). New York: McGraw-Hill.

Cooper, W.J. 1996. Personal discussions with Leo Weitzman. April 12.

Cooper, W.J., T.D. Waite, C.N. Kurucz, M.G. Nickelsen, and K. Lin. 1992. *An overview of the use of high energy electron beam irradiation for the destruction of toxic organic chemicals from water, wastewater, and waters containing solids*. Draft paper provided by Dr. William Cooper, High Voltage Environmental Applications, Inc. 9562 Doral Boulevard, Miami, FL 33178.

Cooper, W.J., E. Cadavid, M.G. Nickelsen, K. Lin, C.N. Kurucz, and T.D. Waite. 1993a. Removing THMs from drinking water using high-energy electron-beam irradiation. *Journal of the American Water Works Association*. September: 106-112.

Cooper, W.J., D.E. Meacham, M.G. Nickelsen, K. Lin, D.B. Ford, C.N. Kurucz and T.D. Waite. 1993b. The removal of tri- (TCE) and tetrachloroethylene (PCE) from aqueous solution using high-energy electrons. *Journal of the Air & Waste Management Association*. 43: 1358-1366.

Cornwell, D.A. 1990. Air stripping and aeration. *Water Quality and Treatment, A Handbook of Community Water Supplies.* American Water Works Association. 4th edition. F.W. Pontius (ed.). New York: McGraw-Hill.

Crain, N. and E.F. Gloyna. 1992. Supercritical water oxidation of 2,4-dichlorophenol and pyridine. Paper presented at the Separations Research Program Fall Conference. The University of Texas at Austin. October 14-15.

Dell'Orco, P. 1991. The separation of particles from supercritical water oxidation effluents. Masters Thesis. Civil Engineering Department, The University of Texas at Austin.

Dell'Orco, P., E.F. Gloyna, and M. Buelow. 1991. The separation of solids from supercritical water. *Supercritical Fluids Engineering.* American Chemical Society.

Dell'Orco, P., L. Li, and E.F. Gloyna. 1993. The separation of particulates from supercritical water oxidation processes. *J. Separation Science and Technology for Energy Applications.* 28(1-3): 625-642.

Dickinson, N.L. and C.P. Welch. 1958. Heat transfer to supercritical water. Transactions of the ASME. 80(3): 746-52.

Electro-Petroleum, Inc. Date unknown. Electrokinetic treatment of contaminated soils, sludges, and lagoons. DOE/CH-9206.

EnviroMetals Technologies, Inc. Undated. The EnviroMetal Process. Technical release. Mailing address: 42 Arrow Road, Guelph, Ontario, Canada, N1K1S6.

Evans, 0. 1990. Estimating innovative treatment technology costs for the SITE program. *Journal of the Air and Waste Management Association.* Volume 40, No. 7. July.

Felmy, A.R., D.C. Girvin, and E.A. Jenne. 1984. MINTEQ-A. A computer program for calculating aqueous geochemical equilibria. USEPA-600/3-84-032. Athens, GA: US EPA.

Focht, R., J. Vogan, and S. F. O'Hannesin. 1996. Field application of reactive iron walls for in-situ degradation of volatile organic compounds in groundwater. *Remediation.* Summer: 81-94.

Foussard, J.N., H. Debellefontain, and J. Besombes-Vailhe. 1989. Efficient elimination of organic liquid wastes: wet air oxidation. *J. Environmental Eng.* 115(2): 367-385.

Garrels, R.M. and C.L. Christ. 1965. Ion exchange and ion sensitive electrodes. *Solutions, Minerals and Equilibria* . San Francisco: Freeman Cooper and Co.

Gehiinger, P., E. Proksch, W. Szinovatz, and H. Eschweiler. 1988. Decomposition of trichloroethylene and tetrachloroethylene in drinking water by a combined radiation/ozone treatment. *Water Research.* 22: 645.

Gill, J.H. and J.M.Quiel. 1993. *Incineration of Hazardous, Toxic, and Mixed Wastes.* Cleveland, OH: North American Mfg. Co.

Gillham, R.W. 1995. Resurgence in research concerning organic transformations enhanced by zero-valent metals and potential applications in remediation of contaminated groundwater. Paper presented to the Division of Environmental Chemistry, American Chemical Society conference. Anaheim, CA. April 2-7.

Gillham, R.W. and F. O'Hannesin. 1994. Enhanced degradation of halogenated aliphatics by zero-valent iron. *Ground Water.* 32(6): 958-967.

Glaze, W.H. 1990. Chemical oxidation. *Water Quality and Treatment, A Handbook of Community Water Supplies.* American Water Works Association. 4th edition. F.W. Pontius (ed.). New York: McGraw-Hill.

Gloyna, E.F. and L. Li. 1993. Supercritical water oxidation: an engineering update. Paper presented at EPRI-NSF Symposium. San Francisco, CA. February 22-24.

Gloyna, E.F. and L. Li. 1995. Supercritical water oxidation research and development updates. *Environmental Progress.* 14(3): 182-192.

Griffith, J. and E.F. Gloyna. 1992. Supercritical water oxidation of *o*-cresol and methyl ethyl ketone. Paper presented at the separations research program Spring Conference. The University of Texas at Austin. April 14-15.

Hamed, J., Y.B. Acar, and R.J. Gale. 1991. *J. Geotechnical Eng.* New York: ASCE.

Haroldsen, B.L., D.Y. Ariizumi, B.E. Mills, B.G. Brown, and D. Griesen. 1996. Transpiring wall supercritical water oxidation reactor salt deposition studies. Sandia Report SAND96-8255•UC-702. September.

Hazlebeck, D. A., K.W. Downey, and A. J. Roberts. 1994. Development of prototype hydrothermal oxidation systems for propellants, chemical agents, and other hazardous wastes. Extended abstract presented at the American Chemical Society. Atlanta GA. September 19-21.

Helfferich, F. 1962. *Ion Exchange.* New York: McGraw-Hill.

Helling, R.K. 1986. Oxidation kinetics of simple compounds in supercritical water: carbon monoxide, ammonia and ethanol. Ph.D. Dissertation. Massachusetts Institute of Technology, Cambridge, Massachusetts.

Helling, R.K. and J.W. Tester. 1987. Oxidation kinetics of carbon monoxide in supercritical water. *J. Energy & Fuel.* 1: 417.

Hoasain, S.U. and C.A. Blaney. 1991. Method for removing polychiorinated dibensodioxins and polychiorinated dibenzofurans from paper mill sludge. US Patent No. 5,075,017. December 24.

Holden, W., A. Marcellino, D. Valic, and A.C. Weedon. 1993. Titanium dioxide mediated photochemical destruction of trichloroethylene vapors in air. *Photocatalytic Purification and Treatment of Water and Air.* D.F. Ollis and H. Al-Ekabi (eds.). Elsevier Science Publishers B.V. pp 393 and 404.

Holgate, H.R., P.A. Webley, J.W. Tester, and R.K. Helling. 1992. Carbon monoxide oxidation in supercritical water: the effects of heat transfer and the water-gas shift reaction on observed kinetics. *Energy and Fuels.* 6:586-97.

Holser, R.A., S.C. McCutcheon, and N.L. Wolfe. Undated-a. Mass transfer effects on the dechlorination of trichloroethene by iron/pyrite fixtures in continuous flow column systems. Submitted for publication. Copy can be obtained from S.C. McCutcheon at Ecosystems Research Division, National Exposure Research Laboratory, U.S. Environmental Protection Agency, Athens, GA 30605.

Holser, R.A., S.C. McCutcheon, and N.L. Wolfe. Undated-b. The application of a rate expression for the reductive dehalogenation of TCE-contaminated groundwater to the design and performance of an in situ reactive zone. Submitted for publication. Copy can be obtained from S.C. McCutcheon at Ecosystems Research Division, National Exposure Research Laboratory, U.S. Environmental Protection Agency, Athens, GA 30605.

Hong, M.S., L. He, B.E. Dale, and K.C. Donnelly. 1995. Genotoxicity profiles and reaction characteristics of potassium polyethylene glycol dehalogenation of wood preserving waste. *Env. Sci. Tech.* 29(3): 702-708.

Hutchenson, K.W., and N.R. Foster. 1995. *Innovations in Supercritical Fluids.* Washington, DC: American Chemical Society.

Hunter, R.J. 1982. *Zeta Potential of Colloid Science.* London: Academic Press.

Idaho National Engineering Laboratory (INEL). 1995. Corrosion investigation of multilayered ceramics and experimental nickel alloys in SCWO process environments. INEL-95/0017.

Jacobs, G.P., H.H. Barner, and A.L. Bourhis. 1992. Utilizatrion of phenolics in the design of the MODAR SCWO reactor. AIChE Annual Meeting. Miami, FL. November 1-6.

James M. Montgomery, Consulting Engineers, Inc. 1985. *Water Treatment Principles and Design*. New York: John Wiley & Sons.

Jeffers, P.M., L.M. Ward, L.M. Woytowitch, and N.L. Wolfe. 1989. Homogeneous hydrolysis rate constants for selected chlorinated methanes, ethanes, ethenes, and propanes. *Env. Science & Tech.* 23(8): 965-969.

Johnson, T.L., M.M. Shcerer, and P.G. Tratnyek. 1996. Kinetics of halogenated organic compound degradation by iron metal. *Environ. Sci. Technol.* 30(8).

Kahn, L.I. and M.S. Alam. 1993. Heavy metal removal from soil by coupled electric-hydraulic gradient. *J. Env. Eng., ASCE.* 120(6): 1524-1543.

Kalen, David V. 1992. Destruction of chemical warfare agents simulants in water solutions by cobalt-60 gamma irradiation. Masters Thesis. Florida International University, Miami, Florida.

Killilea, William. 1996. Personal correspondence. From William Killea of Modar, General Atomics to Leo Weitzman.

Killilea, W.R., K.C. Swallow, and G.T. Hong. 1992. The fate of nitrogen in supercritical waste oxidation. *J. Supercritical Fluids.* 5(1): 72-78.

Kurucz, C.N., T.D. Waite, W.J. Cooper, and M.G. Nickelsen. 1991a. Full-scale electron beam treatment of hazardous wastes — effectiveness and costs. *Proc. 45th Industrial Waste Conference* (May 8-10, 1990). Chelsea, MI: Lewis Publishers.

Kurucz, C.N., T.D. Waite, W.J. Cooper, and M.G. Nickelsen. 1991b. *Advances in Nuclear Science and Technology*. Volume 22. J. Lewins and M. Becker (eds.). New York: Plenum Press.

Kurucz, C.N., T.D.Waite and W.J. Cooper. 1995. The Miami electron beam research facility: a large scale wastewater treatment application. *Radiation Physical Chemistry*. 45(2): 299-308.

Kurucz, C.N., T.D. Waite, W.J. Cooper, and M.G. Nickelsen. 1995. Empirical models for estimating the destruction of toxic organic compounds utilizing electron beam irradiation at full-scale. *Radiation Physical Chemistry*. 45(5): 805-816.

Lageman, R. 1993. Electroreclamation: applications in the Netherlands. *Environ. Sci. & Tech.* 27(13): 2648-2650.

Laschinger, M.N. 1992. Pilot study report milan army anmuntion plant o-line ponds area. Study performed by Andco Environmental Processes, Inc., 595 Commerce Drive Amherst, NY 14228.

Lee, D.S. 1990. Supercritical water of acetamide and acetic acid. PhD Dissertation. Civil Engineering Department, The University of Texas at Austin.

Leith, D. and W. Licht. 1972. The collection efficiency of cyclone type particle collectors: a new theoretical approach. American Institute of Chemical Engineers Symposium Series No. 126. 68:196.

Levenspiel. 1962. *Chemical Reaction Engineeering*. New York: John Wiley & Sons.

Li, L., P. Chen, and E.F. Gloyna. 1991. Generalized kinetic model for wet oxidation of organic compounds. *AIChEJ.* 37(11):1687-97.

Lindgren, E.R., E.D. Mattson, and M.W. Kozak. 1994. Electrokinetic remediation of unsaturated soils. *Emerging Technologies in Hazardous Waste Management IV*. ACS Symposium Series 554. D.W. Tedder and F.G. Pohland (eds.). Washington, DC: American Chemical Society.

Lindgren, E.R., M.W. Kozak, and E.D. Mattson. 1991. Electrokinetic remediation of contaminated soils. Report No.: SAND-91-0726C. Sandia National Labs.

Lyons, T. 1995. Personal communication with T. Lyons, Risk Reduction Engineering Laboratory. US EPA. August.

Lyons, T. 1996. Personal communication with T. Lyons, National Risk Management Research Laboratory. US EPA. January.

Marks, R.E. 1995. Personal communication, with R.E. Marks, Electrokinetics, Inc. August.

Marks, R.E., Y.B. Acar, and R.J. Gale. 1994. In situ remediation of contaminated soils containing hazardous mixed wastes. by bio-electrokinetic remediation and other competitive technologies. *Remediation of Hazardous Waste Contaminated Soils.* D.L. Wise and D.J. Trantolo (eds.). New York: Marcel Dekker, Inc.

Matheson, L.J. and P.G. Tratnyek. 1994. Reductive dehalogenation of chlorinated methanes by iron metal. *Environ. Sci. & Tech.* 28(12): 2045-2053.

Mattson, E.D. and E.R. Lindgren. 1995. Electrokinetic extraction of chromate from unsaturated soils. *Emerging Technologies in Hazardous Waste Management.* ACS Series 607. D.W. Tedder and F.G. Pohland (eds.). Washington, DC: American Chemical Society.

McAdams, W.H., W.E. Kennel, and J.N. Addoms. 1950. Heat transfer to superheated steam at high pressures. Transactions of the ASME. 72(5):421-428.

McCutcheon, S.C. 1996. Personal communication with S.C. McCutcheon. US EPA National Exposure Research Laboratory, Athens, GA. January.

Michna, R. 1990. Heat transfer to water in countercurrent flow within a vertical, concentric-tube supercritical water oxidation reactor. Masters Thesis. Civil Engineering Department, The University of Texas at Austin.

Mills, W.B., D.B. Porcella, M.J. Ungs, S A. Gherini, K.V. Summers, L. Mok, G.L. Rupp, and G.L. Bowie. 1985 (Revised). *Water Quality Assessment: A Screening Procedure for Toxic and Conventional Pollutants in Surface and Ground Water - Part I.* USEPA/600/6-85/002a.

Nickelsen, M.G., W.J. Cooper, C.N. Kurucz, and T.D. Waite. 1992. Removal of benzene and selected alkyl-substituted benzenes from aqueous solution utilizing continuous high-energy electron irradiation. *Environnmental Science and Technology.* 26: 144-152.

O'Hannesin, S.F. and R.W. Gillham. 1992. A permeable reaction wall for in situ degradation of halogenated organic compounds. Paper presented at the 45th Canadian Geotechnical Society Conference. Toronto, Ontario, Canada. October 25-27.

Palmer, C.D. and P.R. Wittbrodt. 1991. Processes affecting the remediation of chromium-contaminated sites. *Environmental Health Perspectives.* 92: 25-40.

Pamukca, S. and J.K. Wittle. 1992. Electrokinetic removal of selected heavy metals from soil. *Environ. Prog.* 11(3): 241-250.

Parks, G.A. 1967. Equilibrium concepts in natural water systems. *Advances in Chemistry Series 67.* Washington, DC: American Chemical Society. p121

Perry, R.H., C.H. Chilton, and S.D. Kirkpatrick. 1963. *Chemical Engineers' Handbook*, New York: McGraw-Hill.

PRC Environmental Management, Inc.. 1994. HVAEA electron beam technology demonstnation final quality assurance project plan. Submitted to US EPA Office of Research and Development (ORD). Cincinnati, OH. September.

Probstein, R.F. 1989. *Physicochemical Hydrodynamics: An Introduction.* Boston and London: Butterworth-Heinemann.

Probstein, R.F. and R.E. Hicks. 1993. Removal of contaminants from soils by electric fields. *Science.* 260: 498-503.

Rofer, C.K. and G.E. Streit. 1989. *Oxidation of Hydrocarbons and Oxyenates in Supercritical Water, Phase II Final Report.* LA-11700-MS, DOE/HWP-90. Los Alamos National Laboratory, Los Alamos, NM.

Rogers, C.J. 1994. Basic chemistry of the ease catalyze decomposition process and its applications nationally and internationally. Presented at the I&EC Special Symposium, American Chemical Society. Atlanta, GA. September 19-21.

Rollans, S., L. Li, and E.F. Gloyna. 1992. Separation of hexavalent and trivalent chromium from supercritical water oxidation effluents. Paper presented at the Industrial & Engineering Chemistry Special Symposium, American Chemical Society. Atlanta, GA. September 21-23.

Rousar, D.C., Marvin F. Young, and Scott Sieger. 1995. Development of Components for Waste Management Systems Using Aerospace Technology. 46th International Astronautical Congress. Oslo, Norway. American Institute of Aeronautics and Astronautics Paper #IAA 95-IAA.1.2.07. October 2-6.

Runnells, D.D. and C. Wahli. 1993. In situ electromigration as a method for removing sulfate, metals, and other contaminants from ground water. *Ground Water Monitoring & Remediation.* Winter: 121-129.

Scherer, M.M. and P.G. Tratnyk. 1995. Dechlorination of carbon tetrachloride by iron metal: effect of reactant concentration. Paper presented to the Division of Environmental Chemistry, American Chemical Society. Anaheim, CA. April 2-7.

Segall, B.A. and C.J. Bruell. 1992. Electroosmotic contaminant removal processes. *J. Env. Eng., ASCE.* 118(1): 84-100.

Senzaki, T. 1991. Removal of chlorinated organic compounds from wastewater by reduction process: III treatment of trichloroethylene with iron powder II. *Kogyo Yosui.* 391: 29-35 (in Japanese).

Senzaki, T. and Y. Kumagai. 1988. Removal of chlorinated organic compounds from wastewater by reduction process: treatment of 1,1,2,2-tetrachloroethane with iron powder. *Kogyo Yosui.* 357: 2-7 (in Japanese).

Senzaki, T. and Y. Kumagai. 1989. Removal of chlorinated organic compounds from wastewater by reduction process: II treatment of trichloroethylene with iron powder. *Kogyo Yosui.* 369: 19-25 (in Japanese).

Shanableh, A.M. 1990. Subcritical and supercritical water oxidation of industrial, excess activated sludge. PhD Dissertation. Civil Engineering Department, The University of Texas at Austin.

Shapiro, A.P. and R.F. Probstein. 1993. Removal of contaminants from saturated clay by electroosmosis. *Environ. Sci. & Tech.* 27(2): 283-291.

Shapiro, A.P., P. Renauld, and R.F. Probstein. 1989. Preliminary studies on the removal of chemical species from saturated porous media by electroosmosis. *Physicochemical Hydrodynamics.* 11(5/6): 785-802.

Shoemaker, S. H., J. F. Greiner, and R. W. Gillham. 1995. Permeable reactive barriers. Chapter 11 of Summary Report from the International Workshop on Containment Technologies for Environmental Remediation Applications, Baltimore, MD. New York: John Wiley & Sons.

Siddiqui, M., G.L. Amy, W.J. Cooper, C.N. Kurucz, T.D. Waite, and M.G. Nickelsen. 1996. Bromate removal by the high energy electron beam process. *Journal of the American Water Works Association.* 88(10): 90-101.

Sivavec, T.M. and D.P. Horney. 1995. Reductive dechlorination of chlorinated ethenes by iron metal. Paper presented to the Division of Environmental Chemistry, American Chemical Society. Anaheim, CA. April 2-7.

Shaw, R.S., T.B. Brill, A.A. Clifford, C.A. Eckert, and E.U. Franck. 1991. Supercritical water: a medium for chemistry. *C & EN*. December 23.

Snoeyink, V.L. 1990. Adsorption of organic compounds. *Water Quality and Treatment, A Handbook of Community Water Supplies.* American Water Works Association. 4th edition. F.W. Pontius (ed.). New York: McGraw-Hill.

Snoeyink, V.L. and D. Jenkins. 1980. *Water Chemistry.* New York: John Wiley & Sons.

Spinks, J.W.T. and R.J. Woods. 1990. *An Introduction to Radiation Chemistry.* New York: John Wiley & Sons.

Spritzer, M. 1993. Personal communication with Michael Spritzer, General Atomics. October.

Starr, R.C. and J.C. Cherry. 1993. Ground water currents. US EPA. USEPA/542/N-93/006. June.

State of Connecticut Department of Environmental Protection (CDEP). 1993. Personal Communication. From A. Lacobucci to US EPA regarding the acute toxicity of hydrogen peroxide to freshwater organisms.

Stradling, J. 1989. A review of the hazards, safety aspects, design concepts, and contamination control considerations relating to the safe use of liquid and gaseous oxygen. Prepared by Jack Stradling at the NASA White Sands Test Facility. March 21.

Tester, J.W., H.R. Holgate, F.J. Armellini, P.A. Webley, W.R. Killilea, G.T. Hong, and H.E. Barner. 1993. Supercritical water oxidation technology: process development and fundamental research. *Emerging Technologies in Hazardous Waste Management III.* ACS Symposium Series 518. D.W. Tedder and F.G. Poland (eds.). American Chemical Society. 0097-6156/93/0518-0035.

Timberlake, D. 1993. Personal communication. Risk Reduction Engineering Laboratory, USEPA. August.

Tongdhamachart, C. 1996. Supercritical water oxidation of anaerobically digested municipal sludge. Doctoral Dissertation. Civil Engineering Department, The University of Texas at Austin.

Tratnyek, P.G. 1996. Putting corrosion to use: remediating contaminated groundwater wuth zerovalent metals. *Chemistry and Industry.* July. pp 499-503.

Turner, M. D. 1993. Supercritical water oxidation of dimethyl methylphosphonate and thiodiglycol. ARPA project sponsored, Doctoral dissertation. University of Texas at Austin. December.

U.S. Department of Energy (DOE). 1988. Radioactive Waste Management Order. DOE Order 5820.2A. September.

US EPA. 1983. Methods for chemical analysis of water and wastes. Environmental Monitoring and Support Laboratory. Cincinnati, OH. US EPA/600/4-79/020. March.

US EPA. 1987a. Alternate Concentration Limit (ACL) Guidance, Part I: ACL Policy and Information Requirements. US EPA/530/SW-87/017.

US EPA. 1 987b. Joint US EPA-Nuclear Regulatory Agency Guidance on Mixed Low-Level Radioactive and Hazardous Waste. Office of Solid Waste and Emergency Response (OSWER) Directives. 9480.00-14 (June 29), 9432.00-2 (January 8), and 9487.00-8 (August).

US EPA. 1988a. *Protocol for a Chemical Treatment Demonstration Plan.* Hazardous Waste Engineering Research Laboratory. Cincinnati, OH. April.

US EPA. 1988b. *CERCLA Compliance with Other Environmental Laws*: Interim Final. OSWER. US EPA/540/G-89/006. August.

US EPA. 1989. *CERCLA Compliance with Other Laws Manual: Part II Clean Air Act and Other Environmental Statutes and State Requirements.* OSWER. US EPA/540/G-89/006. August.

US EPA. 1990. *Test Methods for Evaluating Solid Waste.* Volumes lA-IC. SW-846. 3rd Edition. Update I. OSWER. Washington, DC. November

US EPA. 1992. Electron beam treatment for removal of trichloroethylene and tetrachioroethylene from streams and sludge. *Emerging Technology Bulletin.* BPAli40/F~92/()()9. October.

US EPA. 1993. Electron beam treatment for the removal of bentene and toluene from aqueous streams and sludge. *Emerging Technology Bulletin.* US EPAl540/F-93/302. April.

US EPA. 1994. Electrokinetics Inc. (Electro-Klean™ Electrokinetic soil processing). *Superfund Innovative Technology Evaluation Program; Technology Profiles.* 7th edition. USEPA/540/R-94/526, 172-173.

US EPA. 1995a. *In-Situ Remediation Technology Status Report: Treatment Walls.* USEPA542-K-94-004. April.

US EPA. 1995b. Metal-enhanced abiotic degradation technology. EnviroMetal Technologies, Inc. *Superfund Innovative Technology Evaluation (SITE) Demonstration Bulletin.* In publication.

US EPA. 1995c. Electrokinetics soil processing (Electrokinetics Inc.). *Superfund Innovative Technology Evaluation Program; Emerging Technology Bulletin.* USEPA/540/F-95/504.

US EPA. 1995d. National National Risk Management Research Laboratory (NRML), High Voltage Environmental Applications, Inc. — Electron Beam Technology, Innovative Technology Evaluation Report. June. Draft Report Available from NRML/SITE Program, US EPA, Cincinnati, OH 45268

Ugaz, A., S. Puppala, R.J. Gale, and Y.B. Acar. 1994. Electrokinetic soil processing. Complicating features of electrokinetic remediation of soils and slurries: saturation effects and the role of the cathodic electrolysis. *Chem. Eng. Comm.* 129: 183-200.

Vogan, John. 1996. Personal communication. John Vogan of EnviroMetal Technologies with B. Kim. January 19.

Vogan, J.L., R.W. Gillham, S.F. O'Hannesin, and W.H.Mautulewicz. 1995. Site specific degradation of VOCS in groundwater using zero valent iron. Paper presented to the Division of Environmental Chemistry, American Chemical Society. Anaheim CA. April 2-7.

Vogel, T.M., C.S. Criddle, and P.L. McCarty. 1987. Transformation of halogenated aliphatic compounds. *Env. Science & Technol.* 21(8): 722-736.

Waite, T.D., W.J. Cooper and C.N. Kurucz. 1993. Electron beam systems for treatment of water and slurried soils contaminated with toxic organics and ordinance residues. Paper presented at 17th Annual Army Environmental R&D Symposium and Third USACE Innovative Technology Transfer Workshop. Williamsburg, VA. June 22-24.

Waite, T.D., T. Wang, C.N. Kurucz, and W.J. Cooper. 1996. Electron beam treatment of bisolids: parameters effecting dewaterability enhancement. Submitted to *Journal of Environmental Engineering.* March.

Weber, W.J., Jr. 1972. *Physicochemical Processes for Water Quality Control.* New York: Wiley-Interscience.

Webley, P.A., J.W. Tester, and H.R. Holgate. 1991. Oxidation kinetics of ammonia and ammonia-methanol mixtures in supercritical water in the temperature range 530-700°C at 246 bar. *Ind. Eng. Chem. Res.* 30(8): 1745-1754.

Weismantel, G. 1996. Supercritical water oxidation treats toxic organics in sludge. *Environmental Technology.* 6(5). September-October.

Weitzman, Leo, Kimberly Gray, Frederick K. Kawahara, Robert W. Peters, and John Verbicky. 1994. *Innovative Site Remediation Technology — Chemical Treatment.* Annapolis, MD: American Academy of Environmental Engineers.

White, G.C. 1972. *Handbook of Chlorination for Potable Water, Wastewater, Cooling Water, Industrial Processes, and Swimming Pools.* New York: Van Nostrand Reinhold.

Wightman, T.J. 1981. Studies in supercritical wet air oxidation. Masters Thesis. Chemical Engineering Department, University of California, Berkeley, CA.

Wilson, E.K. 1995. Zero-valent metals provide possible solution to groundwater problems. *Chemical & Engineering News.* July 3: 19-22.

Yamagata, K., K. Nishikawa, S. Hasegawa, T. Fuji, and S. Yshida. 1972. Forced convective heat transfer to supercritical water flowing in tubes. *Inter. J. Heat and Mass Transfer.* 15:2575-93.

Yamane, C. L., S.D. Warner, J.D. Gallinatti, F.S. Szerdy, T.A. Delfino, D.A. Hankins, and J.L. Vogan. 1995. Installation of a subsurface groundwater treatment wall composed of granular zero-valent iron. Paper presented to the Division of Environmental Chemistry, American Chemical Society conference. Anaheim, CA. April 2-7.

Zhang S. and J.F. Rusling. 1995. Dechlorination of PCB on soils and clay by electrolysis in a bicontinuous microemulsion. *Environ. Sci. Technol.* 29: 1195-1199.

C

SUGGESTED READING LIST

Bechtold, J.K. and S.B. Margolis. 1992. The structure of supercritical diffusion flames with arrhenius mass diffusivities. *Combust. Sci. and Tech.* 83(257).

Bergan, N.E., P.B. Butler, and H.A. Dywer. 1991. High pressure thermodynamics in supercritical water oxidation processes. Presented at the 2nd International Symposium on Supercritical Fluids. Boston, MA. May 20-22.

Bramlette, T.T., B.E. Mills, K.R. Hencken, M.E. Brynildson, S.C. Johnston, J.M. Hruby, H.C. Feemster, B.C. Odegard, and M. Modell. 1991. Destruction of DOE/DP surrogate wastes with supercritical water oxidation technology. Sandia Report SAND90-8229.

Brown, M.S. and R.R. Steeper. 1991. CO_2-based thermometry of supercritical water oxidation. *Appl. Spectrosc.* 46(1733).

Butler, P.B., N.E. Bergan, T.T. Bramlette, W.J. Pitz, and C.K. Westbrook. 1991. Oxidation of hazardous waste in supercritical water: a comparison of modeling and experimental results for methanol destruction. WSS/CI 91-8. Presented at the 1991 Spring Meeting of the Western States Section of the Combustion Institute. Boulder, CO. March 17-19.

Chan, J.P.C., C.A. LaJeunesse, and S.F. Rice. 1994. Experimental techniques to determine salt formation and deposition in supercritical water oxidation reactors. Presented at the 1994 International Mechanical Engineering Congress and Exposition. Chicago, IL. November 6-11.

Croiset, E. and S.F. Rice. 1996. Hydrogen peroxide decomposition in supercritical water. Submitted for publication to AIChE Journal. December.

Hanush, R.G., S.F. Rice, T.B. Hunter, and J.D. Aiken. 1996. Operation and performance of the supercritical fluids reactor (SFR). Sandia National Laboratories Report, SAND96-8203. Livermore, CA.

Haroldsen, B.L., D.Y. Ariizumi, B.E. Mills, B.G. Brown, and D. Greisen. 1996. Transpiring wall supercritical water oxidation reactor salt deposition studies. Sandia National Laboratories Report, SAND96-8255. Livermore, CA.

Hunter,T.B., S.F. Rice, and R.G. Hanush. 1996. Raman spectroscopic measurement of oxidation in supercritical water II. Conversion of isopropanol to acetone. *Industrial and Engineering Chemistry Research.* 35: 3984-3990.

Johnston, S.C. and C.E. Tyner. 1990. Thermochemical waste processing technology at Sandia National Laboratories. Sandia Report SAND90-8692.

LaJeunesse, C.A., B.E. Mills, and B.G. Brown. 1994. Supercritical water oxidation of ammonium picrate. Sandia Report, SAND95-8202.

LaJeunesse, C.A., B.L. Haroldsen, S.F. Rice, and B.G. Brown. 1996. Hydrothermal oxidation of navy shipboard excess hazardous materials. Sandia National Laboratories Report. December. In preparation.

LaJeunesse, C.A., S.F. Rice, R.G. Hanush, and J.D. Aiken. 1993. Salt deposition in a supercritical water oxidation reactor (Interim Report). Sandia Report, SAND94-8201.

LaJeunesse, C.A., J.P. Chan, T.N. Raber, D.C. Macmillan, S.F. Rice, and K.L. Tschritter. 1993. Supercritical water oxidation of colored smoke, dye, and pyrotechnic compositions. *Final Report: Pilot Plant Conceptual Design.* Sandia Report, SAND94-8202.

LaJeunesse, C.A., S.F. Rice, J.J. Bartel, M. Kelley, C.A. Seibel, L.G. Hoffa, T.F. Eklund, and B.C. Odegard. 1992. A supercritical water oxidation reactor: the materials evaluation reactor (MER). Sandia Report, SAND91-8623.

Margolis, S.B. and S.C. Johnston. 1989. Multiplicity and stability of supercritical combustion in a nonadiabatic tubular reactor. *Combust. Sci. and Tech.* 65(103).

Margolis, S.B. and S.C. Johnston. 1990. Nonadiabicity, stoichiometry, and mass diffusion effects on supercritical combustion in a tubular reactor. *Symp. (Int.) Combustion (Proc.) 23rd.* Volume 533.

Margolis, S.B. and S.F. Rice. 1991. On completeness of combustion in an isothermal flow reactor. *Combust. Sci. and Tech.* 78(7).

Melius, C.F., N.E. Bergan, and J.E. Shepherd. 1990. Effects of water on combustion kinetics at high pressure. *Symp. (Int.) Combustion (Proc.) 23rd.* Volume 217.

Rice, S.F., C.A. LaJeunesse, and R.R. Steeper. 1996. Optical cells for high-temperature and high-pressure raman spectroscopic applications. December. In preparation.

Rice, S.F., R.R. Steeper, and C.A. LaJeunesse. 1993. Destruction of representative navy wastes using supercritical water oxidation. Sandia Report, SAND94-8203.

Rice, S.F., R.R. Steeper, and C.A. LaJeunesse. 1993. Efficiency of supercritical water oxidation for the destruction of industrial solvent waste. Paper #22.2. Presented at the I&EC Special Symposium of the American Chemical Society. Atlanta, GA. September 27-29.

Rice, S.F., T.B. Hunter, and R.G. Hanush. 1996. Oxidative reactivity of simple alcohols in supercritical water using in situ raman spectroscopy. Presented at the Second International Symposium on Environmental Applications of Advanced Oxidation Technologies. San Francisco, CA. February 28-March 1. Paper to appear in EPRI proceedings.

Rice, S.F., T.B. Hunter, Å.C. Rydén, and R.G. Hanush. 1996. Raman spectroscopic measurement of oxidation in supercritical water I. Conversion of methanol to formaldehyde. *Industrial and Engineering Chemistry Research.* 35: 2161-2171.

Rice, S.F., C.A. LaJeunesse, R.G. Hanush, J.D. Aiken, and S.C. Johnston. 1994. Supercritical water oxidation of colored smoke, dye, and pyrotechnic compositions. Sandia Report, SAND94-8209.

Rice, S.F., R.G. Hanush, T.B. Hunter, R.R. Steeper, J.D. Aiken, E. Croiset, and C.A. LaJeunesse. 1996. Kinetic investigation of the oxidation of naval excess hazardous materials in supercritical water for the design of a transpiration-wall reactor. Sandia National Laboratories Report. December. In press.

Schmitt, R.G., P.B. Butler, N.E. Bergan, W.J. Pitz, and C.K. Westbrook. 1991. Destruction of hazardous waste in supercritical water II: a study of high-pressure methanol oxidation kinetics. Presented at the Fall Meeting of the Western States Section of the Combustion Institute. Los Angeles, CA. October 13-15.

Steeper, R.R. and S.F. Rice. 1993. Supercritical water oxidation of hazardous wastes. AIAA-93-0810. 31st Aerospace Sciences Meeting. Reno, NV. January 11-14.

Steeper, R.R. and S.F. Rice. 1994. Optical monitoring of the oxidation of methane in supercritical water. Presented at the Spring Meeting of the Western States Section of the Combustion Institute. Davis, CA. March 21-22.

Steeper, R.R. and S.F. Rice. 1994. Optical monitoring of the oxidation of methane in supercritical water. Presented at the Twelfth International Conference on the Properties of Water and Steam. Orlando, FL. September 11-16.

Steeper, R.R. and S.F. Rice. 1995. Optical monitoring of the oxidation of methane in supercritical water. *Physical Chemistry of Aqueous Systems.* H.J., White, J.V. Sengers, D.B. Neumann, and J.C. Bellows (eds.). New York: Begell House. p 652.

Steeper, R.R. and S.F. Rice. 1996. Kinetics measurements of methane in supercritical water. *Journal of Physical Chemistry.* 100: 184-189.

Steeper, R.R., J.D. Aiken, and S.F. Rice. 1996. Kinetics of the water-gas shift reaction in supercritical water and high pressure steam. December. In preparation.

Steeper, R.R., S.F. Rice, M.S. Brown, and S.C. Johnston. 1992. Methane and methanol diffusion flames in supercritical water. *J. Supercritical Fluids.* 5(262).

Steeper, R.R., S.F. Rice, M.S. Brown, and S.C. Johnston. 1992. Methane and methanol diffusion flames in supercritical water. Sandia Report SAND 92-8474.

THE WASTECH® MONOGRAPH SERIES (PHASE II) ON INNOVATIVE SITE REMEDIATION TECHNOLOGY: DESIGN AND APPLICATION

This seven-book series focusing on the design and application of innovative site remediation technologies follows an earlier series (Phase I, 1994-1995) which cover the process descriptions, evaluations, and limitations of these same technologies. The success of that series of publications suggested that this Phase II series be developed for practitioners in need of design information and applications, including case studies.

WASTECH® is a multiorganization effort which joins in partnership the Air and Waste Management Association, the American Institute of Chemical Engineers, the American Society of Civil Engineers, the American Society of Mechanical Engineers, the Hazardous Waste Action Coalition, the Society for Industrial Microbiology, the Soil Science Society of America, and the Water Environment Federation, together with the American Academy of Environmental Engineers, the U.S. Environmental Protection Agency, the U.S. Department of Defense, and the U.S. Department of Energy.

A Steering Committee composed of highly respected members of each participating organization with expertise in remediation technology formulated and guided both phases, with project management and support provided by the Academy. Each monograph was prepared by a Task Group of recognized experts. The manuscripts were subjected to extensive peer reviews prior to publication. This Design and Application Series includes:

Vol 1 - Bioremediation

Principal authors: **R. Ryan Dupont, Ph.D.**, *Chair*, Utah State University; **Clifford J. Bruell, Ph.D.**, University of Massachusetts; **Douglas C. Downey**, Parsons Engineering Science; **Scott G. Huling**, USEPA; **Michael C. Marley, Ph.D.**, Environgen, Inc.; **Robert D. Norris, Ph.D.**, Eckenfelder, Inc.; **Bruce Pivetz**, USEPA.

Vol 2 - Chemical Treatment

Principal authors: **Leo Weitzman, Ph.D.**, LVW Associates, *Chair*; **Irvin A. Jefcoat, Ph.D.**, University of Alabama; **Byung R. Kim, Ph.D.**, Ford Research Laboratory.

Vol 3 - Liquid Extraction Technologies: Soil Washing/Soil Flushing/Solvent Chemical

Principal authors: **Michael J. Mann, P.E., DEE**, Alternative Remedial Technologies, Inc., *Chair*; **Richard J. Ayen, Ph.D.**, Waste Management Inc.; **Lorne G. Everett, Ph.D.**, Geraghty & Miller, Inc.; **Dirk Gombert II, P.E.**, LIFCO; **Mark Meckes**, USEPA; **Chester R. McKee, Ph.D.**, In-Situ, Inc.; **Richard P. Traver, P.E.**, Bergmann USA; **Phillip D. Walling, Jr., P.E.**, E. I. DuPont Co. Inc.

Vol 4 - Stabilization/Solidification

Principal authors: **Paul D. Kalb**, Brookhaven National Laboratory, *Chair*; **Jesse R. Conner**, Rust Remedial Services, Inc.; **John L. Mayberry**, SAIC; **Bhavesh R. Patel**, Brookhaven National Laboratory; **Joseph M. Perez, Jr.**, Battelle Pacific Northwest; **Russell L. Treat**, Foster Wheeler Environmental Corp.

Vol 5 - Thermal Desorption

Principal authors: **William L. Troxler, P.E.**, Focus Environmental Inc., *Chair*; **Edward S. Alperin**, IT Corporation; **Paul R. de Percin**, USEPA; **Joseph H. Hutton, P.E.**, Canonie Environmental Services, Inc.; **JoAnn S. Lighty, Ph.D.**, University of Utah; **Carl R. Palmer, P.E.**, Rust Remedial Services, Inc.

Vol 6 - Thermal Destruction

Principal authors: **Francis W. Holm, Ph.D.**, SAIC, *Chair*; **Carl R. Cooley**, Department of Energy; **James J. Cudahy, P.E.**, Focus Environmental Inc.; **Clyde R. Dempsey, P.E.**, USEPA; **John P. Longwell, Sc.D.**, Massachusetts Insititute of Technology; **Richard S. Magee, Sc.D., P.E., DEE**, New Jersey Institute of Technology; **Walter G. May, Sc.D.**, University of Illinois.

Vol 7 - Vapor Extraction and Air Sparging

Principal authors: **Timothy B. Holbrook, P.E.**, Camp Dresser & McKee, *Chair*; **David H. Bass, Sc.D.**, Groundwater Technology, Inc.; **Paul M. Boersma**, CH2M Hill; **Dominic C. DiGuilio**, University of Arizona; **John J. Eisenbeis, Ph.D.**, Camp Dresser & McKee; **Neil J. Hutzler, Ph.D.**, Michigan Technological University; **Eric P. Roberts, P.E.**, ICF Kaiser Engineers, Inc.

The monographs for both the Phase I and Phase II series may be purchased from the American Academy of Environmental Engineers®, 130 Holiday Court, Suite 100, Annapolis, MD, 21401; Phone: 410-266-3390, Fax: 410-266-7653, E-mail: aaee@ea.net